Tourism Planning and Development in South Asia

TOURISM PLANNING AND DEVELOPMENT IN SOUTH ASIA

Edited by

Dimitrios Stylidis

Middlesex University London, UK

Boopen Seetanah

University of Mauritius, Mauritius

CABI

CABI is a trading name of CAB International

CABI
Nosworthy Way
Wallingford
Oxfordshire OX10 8DE
UK

Tel: +44 (0)1491 832111
Fax: +44 (0)1491 833508
E-mail: info@cabi.org
Website: www.cabi.org

CABI
WeWork
One Lincoln St
24th Floor
Boston, MA 02111
USA

Tel: +1 (617)682-9015
E-mail: cabi-nao@cabi.org

A catalogue record for this book is available from the British Library, London, UK.

Library of Congress Cataloging-in-Publication Data

Names: Seetanah, Boopen, editor. | Stylidis, Dimitrios, editor.
Title: Tourism planning and development in South Asia / Dimitrios Stylidis, Middlesex University London, U.K., Boopen Seetanah, University of Mauritius, Mauritius.
Description: Boston : CAB International, [2021] | Series: CABI Regional Tourism series | Includes bibliographical references and index. | Summary: "Presents key issues of tourism planning and development in India, Sri Lanka, Pakistan, Bangladesh, Nepal, Maldives, Afghanistan, and Bhutan. Demonstrates experiences, challenges, successful and unsuccessful stories and specific cases of countries within South Asia. Written by scholars with extensive research experience on tourism in these countries"-- Provided by publisher.
Identifiers: LCCN 2020045909 (print) | LCCN 2020045910 (ebook) | ISBN 9781789246698 (Hardback) | ISBN 9781789246704 (eBook) | ISBN 9781789246711 (ePub)
Subjects: LCSH: Tourism--South Asia.
Classification: LCC G155.S585 T69 2021 (print) | LCC G155.S585 (ebook) | DDC 915.40068/4--dc23
LC record available at https://lccn.loc.gov/2020045909
LC ebook record available at https://lccn.loc.gov/2020045910

References to Internet websites (URLs) were accurate at the time of writing.

ISBN-13: 9781789246698 (hardback)
9781789246704 (ePDF)
9781789246711 (ePub)

DOI: 10.1079/9781789246698.0000

Commissioning Editor: Claire Parfitt
Editorial Assistant: Lauren Davies
Production Editor: Shankari Wilford

Typeset by SPi, Pondicherry, India
Printed and bound in the UK by Severn, Gloucester

CABI Regional Tourism Series

BY

Konstantinos Andriotis, Middlesex University London

and

Dimitrios Stylidis, Middlesex University London

Tourism literature includes a plethora of books, conference and journal papers that address various issues of tourism in different regions and sub-regions of the world. However, up to now there has been a lack of a collection of contribution-based books with a regional and sub-regional focus. In order to complement the existing tourism literature with new or emerging topics of particular interest, it is the aim of the **CABI Regional Tourism Series** to offer a collection of books with comparative and individual case studies from at least eight regions and sub-regions of the world. These books will come under the following titles, each of which focuses on some of the countries or country clusters: Tourism in Western Europe; Tourism in Eastern Europe; Tourism in East and Southeast Asia; Tourism in the Middle East; Tourism in Sub-Saharan Africa; Tourism in the Caribbean; Tourism in Latin America; Tourism in Pacific Oceania.

Suggested topics:

- Tourism Development and Planning
- Tourism Economics and Finance
- Tourism Marketing, Management and Information Technology

To discuss a proposal for a book in this series, please contact Konstantinos Andriotis at k.andriotis@mdx.ac.uk or Dimitrios Stylidis at d.stylidis@mdx.ac.uk.

Brief summary of careers

Konstantinos Andriotis is a Professor in Tourism at Middlesex University London and in the past, he was Associate Professor and Head of the Hotel and Tourism Management Department at the Cyprus University of Technology. He holds an MSc in International Hospitality Management (University of Strathclyde, 1995), a PhD in Tourism Development and Planning (Bournemouth University, 2000) and a Postdoc in Tourism Marketing (Hellenic Open University, 2006). He edits the *International Journal of Tourism Policy* and the *Journal of Qualitative Research in Tourism*. He is a member of the Editorial Board of more than 20 journals and Associate Editor of the leading tourism journal *Annals of Tourism Research*. He is author of seven books and over 45 peer-reviewed articles and book chapters. His work has appeared in leading academic journals, such as *Annals of Tourism Research*, *Journal of Travel Research*, *Tourism Management* and *Journal of Sustainable Tourism*. He has an h-index of 29 and more than 3700 citations. He has acted as Chairman and member of the organizing and scientific committees of various international conferences and he has given keynote speeches in several conferences. His research interests are in tourism development and planning, alternative forms of tourism, tourism experience and degrowth.

Dimitrios Stylidis is a Senior Lecturer in Tourism at Middlesex University London. Prior to joining Middlesex, Dimitrios worked for 2 years as a Postdoc Researcher at Ben-Gurion University of the Negev, Israel. He holds a PhD and an MSc in Tourism Management from the University of Surrey, UK. Dimitrios is the Research Notes Editor of the *International Journal of Tourism Policy*. His work has appeared in leading academic journals including the *Annals of Tourism Research*, *Journal of Travel Research* and *Tourism Management*. His current research interests include place image, tourism impacts and residents' attitudes towards tourism development.

Contents

** = corresponding author*

Contributors

Glen Croy
glen.croy@monash.edu
Department of Management, Monash University, Building N Level 7, 900 Dandenong Road, Caulfield East, Victoria 3145, Australia

Tek B. Dangi
tek.dangi@maine.edu
Department of Exercise, Health, and Sport Sciences, University of Southern Maine, 319 Masterton Hall, 71 Bedford Street, PO Box 9300, Portland, ME 04104-9300, USA

Terry DeLacy
terry.delacy@vu.edu.au
Research Institute of Sustainable Industries and Liveable Cities, Victoria University, Footscray Park Campus, 70–104 Ballarat Road, Footscray, Victoria 3011, Australia

Sheereen Fauzel
s.fauzel@uom.ac.mu
Department of Finance and Accounting, Faculty of Law and Management, University of Mauritius, Réduit 80837, Mauritius

Laxman Gautam
laxmangautam99@gmail.com
Nepal Tourism Board, Pradarshani Marg, Kathmandu 44617, Nepal

Anu Treesa George
anu.george@live.vu.edu.au
Research Institute of Sustainable Industries and Liveable Cities, Victoria University, Footscray Park Campus, 70–104 Ballarat Road, Footscray, Victoria 3011, Australia

Azizul Hassan
azizulhassan00@gmail.com
Tourism Consultants Network, The Tourism Society, 3 Port House, Square Rigger Row, London SW11 3TY, UK

Zameelah Khan Jaffur
zam.kjaffur@gmail.com
Department of Finance and Accounting, Faculty of Law and Management, University of Mauritius, Réduit 80837, Mauritius

Min Jiang
Min.Jiang@whittlesea.vic.gov.au
City of Whittlesea, 25 Ferres Boulevard, South Morang, Victoria 3752, Australia

Birendra KC
Birendra.Kc@unt.edu
Department of Hospitality and Tourism Management, University of North Texas, 1155 Union Circle #311100, Denton, TX 76203-5017, USA

Abdul Wassay Najimi
aw_najimi@yahoo.com
Formerly at Aga Khan Trust for Culture, Qala-i Moen Afzal Khan, Gozargah Street, District 3, PO Box (35-005), Central Post Office (1001), Kabul, Afghanistan

Frank Niranjan
niranjanfr_03@yahoo.com
Sri Lanka Council for Agricultural Research Policy, Ministry of Agriculture, 114/9 Wijerama Mawatha, Colombo 00700, Sri Lanka

Stephen Pratt
stephen.pratt@usp.ac.fj
School of Tourism and Hospitality Management, The University of the South Pacific, Laucala Campus, Suva, Fiji

Rajasundram Sathiendrakumar
rsathiendrakumar@sheridan.edu.au
Sheridan Institute of Higher Education, Suite 18, 7 Aberdeen Street, Perth, WA 6000, Australia

Boopen Seetanah
b.seetanah@uom.ac.mu
Department of Finance and Accounting, Faculty of Law and Management, University of Mauritius, Réduit 80837, Mauritius

Dimitrios Stylidis
d.stylidis@mdx.ac.uk
Department of Marketing, Branding and Tourism, The Business School, Middlesex University London, The Burroughs, London NW4 4BT, UK

Wantanee Suntikul
w.suntikul@polyu.edu.hk
School of Hotel and Tourism Management, The Hong Kong Polytechnic University, 17 Science Museum Road, TST East, Kowloon, Hong Kong

Contributors' Biographies

Glen Croy is an Associate Professor in the Department of Management, Monash Business School. His teaching and research interests are in tourism, international business and higher education. Three themes emerge from Glen's research. First, demonstrating the socialized 'truth' that films dramatically affect tourist demand is exceptional, although there are many varied and subtle tourism influences. Second, arguing that communities need to be central to tourism and international business decisions, and their achieved desires should be determinants of success. Finally, investigating student team outcomes in regard to the influence of learning approaches and interventions.

Tek B. Dangi, PhD, is an Assistant Professor of Recreation and Leisure Studies in the Department of Exercise, Health and Sport Sciences, University of Southern Maine, USA. Prior to completing a PhD in Recreation, Park and Tourism Sciences from Texas A&M University, Tek accumulated professional experience in tourism planning, management and marketing while serving as a section officer at the Ministry of Culture, Tourism and Civil Aviation, Government of Nepal, and later as CEO of the Nepal Tourism Board. Tek's research interests include tourism planning, sustainable tourism, community-based tourism/ecotourism, and positive youth development (PYD) specifically from Nepal and the USA.

Terry DeLacy is a Professor in Sustainable Tourism in the Research Institute of Sustainable Industries and Liveable Cities, Victoria University, Australia. Previously DeLacy was Director of the Australian National Sustainable Tourism Cooperative Research Centre and Professor of Environmental Policy at the University of Queensland. His research interests are in natural resources and sustainable tourism, most recently focusing on transformation of the tourism sector into the emerging green economy.

Sheereen Fauzel is a Senior Lecturer at the University of Mauritius. Having completed a PhD in International Economics from the University of Mauritius, a Master's degree in Banking and Finance and a Bachelor's degree in Economics and Finance, her areas of expertise are international economics, development economics, tourism economics and related areas. She has participated in international conferences and has publications in notable international business, economics and tourism journals. She is a reviewer for a number of refereed journals including *Current Issues in Tourism* and *Journal of Hospitality Marketing & Management*, among others.

Laxman Gautam, MPhil, is Senior Manager at Nepal Tourism Board and Expert Advisor to the Minister for Culture, Tourism and Civil Aviation, Government of Nepal. He is also a Visiting Lecturer at Tribhuvan University affiliated colleges. His research works are related to sustainable tourism, community-based tourism and rural tourism in the context of Nepal. Laxman has worked in marketing, promotion, media and publicity management of tourism activities of Nepal.

Anu Treesa George is a PhD candidate at the Research Institute of Sustainable Industries and Liveable Cities at Victoria University, Australia. Her research interest is in responsible tourism, hospitality and events industries, green growth strategies and community well-being. Her research focuses on mitigating sustainability challenges by developing a practical framework which combines green growth concepts with responsible tourism to enhance sustainable practices in the tourism industry.

Azizul Hassan is a member of the Tourism Consultants Network of the UK Tourism Society. He is the editor of *Tourism Marketing in Bangladesh: An Introduction* (2020) and is co-editor of *Tourism Events in Asia: Marketing and Development* (2018).

Zameelah Khan Jaffur holds a PhD in Applied Financial Modelling and a BSc (Hons) in Mathematics from the University of Mauritius. She has participated in some international conferences and published in journals such as *Applied Economics Letters*, *Journal of Travel Research* and *Economics Bulletin*. She has also acted as a reviewer for *Tourism Review* and *International Journal of Finance and Economics*. She has research interests in financial modelling, applied econometrics, meta-analysis, exchange rates and related areas.

Min Jiang is a Sustainability Planning and Strategy Officer at the City of Whittlesea, Victoria, Australia, leading the council's climate change strategy development and action planning. Prior to her current local government position, Dr Jiang worked as a Senior Research Fellow at the University of Melbourne and Victoria University. As an environmental lawyer by training, Min's research interests include climate change adaptation, the green economy, sustainable tourism and water governance. She is the author of *Towards Tradable Water Rights: Water Law and Policy Reform in China* (Springer, 2018).

Birendra KC, PhD, is an Assistant Professor in the Department of Hospitality and Tourism Management at the University of North Texas. He is also a Visiting and Affiliate Professor of International Sustainable Tourism for the Tropical Agricultural Research and Higher Education Center (CATIE) in Costa Rica. His research focus is on nature-based tourism, tourism policy and planning, tourism micro-entrepreneurship, community-based tourism and protected area management.

Abdul Wassay Najimi, PhD, is a former teacher of architecture at Kabul University and a conservation architect/educator, educated in Denmark; working from 2002 to 2020 with Aga Khan Trust for Culture (AKTC) in Kabul. He has been involved in field studies, training and projects in architectural conservation with AKTC; work and research with various aid and research organizations (including Danish, Norwegian and Swedish) and the United Nations Educational, Scientific and Cultural Organization (UNESCO) in Afghanistan. He has published articles on Afghanistan architecture and built heritage, is author of the book *Herat: The Islamic City – A Study in Urban Conservation* (1988) and co-author of the book *The Ghurid Portal of the Friday Mosque of Herat, Afghanistan: Conservation of a Historical Monument* (2015).

Frank Niranjan is the Deputy Director (Research Management) in the Sri Lanka Council for Agricultural Research Policy, Ministry of Agriculture. In the role, he has three core responsibilities including, first, managing agricultural research based on the National Agricultural Research Policy and Strategy 2018–2027 and National Research Priorities across 12 subject areas. This involves developing the annual Agricultural Research Status Report of the National Agricultural Research System. Second, monitoring and managing the impact of agricultural research on farmers. Finally, assessing policy-related research undertaken by the Ministry in modernizing agriculture in Sri Lanka.

Stephen Pratt is Professor of the School of Tourism and Hospitality Management at the University of the South Pacific. He was appointed to this position in September 2018. Previously he was at the School of Hotel and Tourism Management at The Hong Kong Polytechnic University from January 2013. He completed his PhD at the University of Nottingham, UK. He obtained a Bachelor of Economics and Master of Economics from the University of Sydney, Australia. His research interests include sustainable tourism development, tourism in small island states and tourist (mis)behaviour.

Rajasundram Sathiendrakumar is Professor at Sheridan Institute of Higher Education in Perth, Western Australia. He graduated in Agriculture with honours from The University of Peradeniya, Ceylon in 1971, is a recipient of a British Council scholarship in Agricultural Economics (1974–1976) and received his Master's degrees at Manchester University, UK. In 1984 he received a scholarship from the University of Newcastle, New South Wales, Australia and was awarded his PhD in 1987. He has held offices by working as a consultant for international organizations including the International Labour Organization (ILO) and the Asian Development Bank (ADB). He has authored over 70 publications, including book chapters, refereed journal articles and conference papers. Rajasundram has been a Visiting Fellow at the NUS University of Dundee, University of Strathclyde, University of Birmingham, University of the South Pacific and the University of Western Sydney.

Boopen Seetanah is a Professor in International Economics and Finance at the University of Mauritius and the World Trade Organization (WTO) Chairs Programme Co-Chair. He is also the director of research at the International Centre for Sustainable Tourism and Hospitality (ICSTH) at the University of Mauritius. Boopen's research interest is in international tourism and trade, international finance and applied economics. He is/has been a consultant for a number of international organizations such as the World Bank, the United Nations Environment Programme (UNEP), the Common Market for Eastern and Southern Africa (COMESA), the Regional Multidisciplinary Centre of Excellence (RMCE) Mauritius, the United Nations Development Programme (UNDP) and the Asian Development Bank (ADB), among others.

Dimitrios Stylidis is a Senior Lecturer in Tourism at Middlesex University London. Prior to this, he was a Postdoc Researcher and Visiting Lecturer at the Ben-Gurion University of the Negev, Israel. Dimitrios holds an MSc and PhD in tourism from the University of Surrey, UK. His current research interests include place image and marketing, tourist destination image and behaviour, tourism impacts and residents' attitudes towards tourism development. His work has been published in several journals including *Tourism Management, Annals of Tourism Research, Journal of Travel Research, Journal of Sustainable Tourism* and *Journal of Travel & Tourism Marketing*.

Wantanee Suntikul is an Independent Researcher based in Thailand and the USA. Her core research interests and expertise are in cultural tourism destinations, political and social aspects of tourism development, religion tourism, heritage tourism, dark tourism, and gastrodiplomacy and tourism. She works on projects in Thailand, Vietnam, Laos, China, Hong Kong, Macao and Bhutan. Dr Suntikul's recent books include: *Tourism and Political Change, Tourism and Political Change* (2nd edn), *Tourism and War, Tourism and Religion: Issues and Implication*. She is also Joint Editor-in-Chief of the journal *Tourism, Culture & Communication*.

Introduction

Boopen Seetanah*

University of Mauritius, Mauritius

South Asia is a unique region of the world and is bound by a common culture, values and history in addition to quite similar social and economic structures (Sobhan, 1999; Dubey, 2007). This sub-region consists of the following eight countries: Afghanistan, Bangladesh, Bhutan, India, Maldives, Nepal, Pakistan and Sri Lanka. With one-fifth of the world's population and only 3.3% of the land mass, it remains among the most densely populated regions globally. The region, with a sizeable contiguous land mass, can boast to have diverse physical characteristics including mangrove forests and grasslands, deserts and swamplands, picturesque mountain ranges (including the Karakorum, the Hindu Kush and the great Himalayas), oceans and coastal areas. Its endowment of natural resources such as rivers, waterfalls and unique scenic beauty together with its contrasting and varying climatic conditions makes this subcontinent quite an appealing place to visit (Rasul and Manandhar, 2009). Moreover, South Asia possesses a uniquely rich and attractive biological and cultural diversity given the unifying presence of Indian, Chinese and Islamic cultures. With its renowned centuries-old civilizations, breathtaking historic sites and other archaeological monuments in addition to its uniquely hospitable people, the subcontinent is poised to be an attractive place for international and particularly intra-regional tourists. This destination has become increasingly appealing for spiritual tourism, with its yoga and meditation retreat centres, traditional healing, Himalayan spas, and famous Muslim mosques and Hindu and Buddhist temples.

It is noteworthy, however, that the economic base of this subcontinent has remained weak, with a below-average gross domestic product (GDP) per capita. Indeed, despite registering slight growth in the manufacturing sector recently,

*E-mail: b.seetanah@uom.ac.mu

 Tourism Planning and Development in South Asia (eds D. Stylidis and B. Seetanah)
DOI: 10.1079/9781789246698.000i

the primary sector, including forestry, agriculture, fisheries and livestock, remains the dominant element of employment, contribution to GDP and foreign exchange earnings. Although South Asia has registered a decent increase in its per capita gross national income since 1970 (from US$118 in 1970 to US$2019 in 2019), it still remains well below the global average of US$11,570 and also that of East Asia (excluding high-income countries), which stood at US$8300, with only Sub-Saharan Africa ranking after it with a gross national income of US$1550 per capita, in 2019.[1] Despite possessing unique and valuable tourist resources and attractions, the potential of tourism in South Asia has remained largely unrealized for several reasons, as discussed in the chapters to follow. The countries of this sub-region have tremendous potential for tourism development and demand, which is already growing at a relatively fast rate of around 5% (World Travel & Tourism Council (WTTC), 2020). Although tourism share in this region remains quite small compared with other regions of the world, interestingly, tourist arrivals in each of the South Asian countries (and the subcontinent as a whole) have been rising continuously since the 1990s (see Table I.1).

The total GDP contribution of the travel and tourism sector to the region amounted to US$234 billion, representing 6.6% of the region's GDP (WTTC, 2020). Table 1.2 shows the pre-COVID-19 total (direct and indirect) contribution of travel and tourism to GDP, the contribution of the sector's employment to total employment, and the sector's percentage contribution to total exports for each country of this region.

It is noteworthy that the South Asia region still remains relatively far behind the North East Asia and South East Asia regions, for which the travel and tourism industry accounts for around 10% and 12% of the economy, respectively, with the Caribbean region standing at 14%. However, South Asia is still relatively highly dependent on travel and tourism in general (with a varying relative tourism economic dependency), especially as a generator of jobs (estimated at 47.7 million in 2019). Moreover, the region's full potential with respect to

Table I.1. Tourist arrivals in South Asia, by country[a], 1995–2019 (source: author's compilation based on data from World Development Indicators (WDI) and individual countries' statistical offices).

	1995	2000	2010	2015	2019
Maldives	315,000	457,000	792,000	1,234,000	1,700,000
Sri Lanka	403,000	400,000	654,000	1,798,000	1,900,000
Bhutan	4,800	7,600	41,000	155,000	208,457
India	2,124,000	2,649,000	5,776,000	13,284,000	10,700,000
Nepal	363,000	465,000	603,000	790,000[c]	1,200,000
Pakistan	423,800[b]	557,000	907,000	1,247,000	1,800,000
Bangladesh	156,000	199,000	303,000	653,000	1,026,000[d]

[a]Figures are not available for Afghanistan.
[b]Figure for year 1990.
[c]Figure for year 2014.
[d]Figure for year 2018.

Table I.2. Contribution of travel and tourism to GDP, employment and total exports in South Asia, by country[a], 2019 (source: author's compilation based on data from WTTC, 2020).

	Contribution of travel and tourism to GDP (%)	Contribution of travel and tourism to employment (%)	International visitor impact on total exports (%)
Maldives	56.6	59.6	79.0
Sri Lanka	10.3	11.0	24.6
Bhutan	9.0	34.0	–
India	6.8	8.0	5.6
Nepal	6.7	6.9	30.8
Pakistan	6.0	6.2	3.1
Bangladesh	3.0	3.0	0.7

[a]Figures are not available for Afghanistan.

tourism development is yet to be tapped and a number of challenges as well as opportunities exist that are documented in this edited volume.

This book is timely and aims to discuss tourism planning and development in South Asia. It supplements the relatively dwarf literature on tourism planning, development and impact in this subcontinent and focuses on the issues and challenges the sector is facing while also emphasizing the opportunities for a number of countries. Chapter 1, by Anu Treesa George, Min Jiang and Terry DeLacy, explores the impacts of community participation and stakeholders' involvement on tourism planning and management in India. Frank Niranjan and Glen Croy in Chapter 2 investigate farmers' perceptions of agritourism potential in Sri Lanka. In Chapter 3, Tek B. Dangi, Birendra KC and Laxman Gautam reflect on tourism planning and development in Nepal, while Chapter 4, authored by Abdul Wassay Najimi, focuses on tourism development in Afghanistan. Stephen Pratt and Wantanee Suntikul in Chapter 5 explore how tourism planning and policy making in Bhutan promote, reinforce or constrain sustainability. Chapter 6, authored by Azizul Hassan, critically analyses tourism policy planning in Bangladesh. In Chapter 7, Sheereen Fauzel, Zameelah Khan Jaffur and Boopen Seetanah identify a series of challenges Pakistan's tourism industry is facing, while Chapter 8 by Rajasundram Sathiendrakumar, Zameelah Khan Jaffur and Boopen Seetanah focuses on the importance of tourism in Maldives. Policy implications are also presented for each of these eight country case studies.

While the chapters of this book were written in a pre-COVID-19 context, it is noteworthy that the pandemic is affecting a significant amount of travel and tourism jobs across this subcontinent, particularly those working in the informal sector. The travel and tourism sector alone is projected to result in an over US$50 billion loss in GDP as a result of the crisis (Twining Ward and McComb, 2020) despite authorities' responses with emergency programmes to support firms and save jobs. South Asia's rich endowment of natural and cultural assets together with its traditionally strong domestic and intra-regional tourism market implies that the region still has good opportunities to seize and can remain relatively competitive in the global tourism industry, despite the sanitary crisis.

Note

[1] Following the outbreak and spread of COVID-19, South Asia's economic outlook is dire and the region has already experienced its worst economic performance of the last decades, with temporary contractions in all eight countries. National output in the region is projected to contract by around 3% in 2020 following COVID-19 owing to sanitary lockdowns and other mitigation measures, alongside a fall in consumption and private investment due to uncertainty (https://www.worldbank.org/en/region/sar/overview, accessed 1 October 2020).

References

Dubey, M. (2007) SAARC and South Asian economic integration. *Economic and Political Weekly* 7, 1238–1241.

Rasul, G. and Manandhar, P. (2009) Prospects and problems in promoting tourism in South Asia: a regional perspective. *South Asia Economy Journal* 10(1), 187–207.

Sobhan, R. (1999) *Transforming Eastern South Asia: Building Growth Zones for Economic Cooperation*. University Press Limited, Dhaka.

Twining Ward, L. and McComb, J.F. (2020) *COVID-19 and Tourism in South Asia: Opportunities for Sustainable Regional Outcomes*. World Bank, Washington, DC. Available at: https://openknowledge.worldbank.org/handle/10986/34050 (accessed 1 October 2020).

WTTC (2020) Travel and Tourism Economic Impact 2019, WTTC 2019 Annual Report. World Travel & Tourism Council. Available at: https://www.wttc.org/-/media/files/reports/economic-impact-research/regions-2019/world2019.pdf (accessed 28 September 2020).

1

A Case Study on Impacts of Community Participation in Tourism Planning and Destination Management in Kerala, India

ANU TREESA GEORGE[1]*, MIN JIANG[2] AND TERRY DELACY[1]

[1]*Victoria University, Australia;* [2]*City of Whittlesea, Australia*

Introduction

Recently, destination design has emerged as a prominent area in tourism research, with the literature focusing on exploring various concepts, perceptions and factors related to destination management (Dredge, 1999; Matsumoto *et al.*, 2012; Fesenmaier and Xiang, 2017). Destination design can contribute to socio-economic improvements and environmental sustainability in tourist destinations (Dredge, 1999; Clark *et al.*, 2009; Paget *et al.*, 2010). Design science in tourism is explicitly focused on developing new artefacts for enabling tourism managers to create innovative processes, systems and places (Fesenmaier and Xiang, 2017). Most designers have outlined the primary stages of design in tourism, which begins with understanding and discovery, progressing to proposing solutions, and finally to evaluation and redesign (Fesenmaier and Xiang, 2017). The factors considered in a destination design are those that 'can enhance the appeal of the core resources and attractors, strengthen the quality and effectiveness of the supporting factors and resources and best adapt to the constraints imposed by the situational conditions' (Crouch and Ritchie, 1999, p. 149). Destination design and management of resources are complicated, as expectations and impacts vary from one destination to the other. The smart design acts as a pathway to relate and include all the internal and external participants in the tourism development process. The best outcome can be generated through constant, dynamic, innovative methods and engagement with all stakeholders that empower communities for sustainable development in a destination (Schaffers *et al.*, 2011).

*E-mail: anu.george@live.vu.edu.au

 Tourism Planning and Development in South Asia
(eds D. Stylidis and B. Seetanah)
DOI: 10.1079/9781789246698.0001

The participatory design and bottom-up approach help to explore the various socio-cultural and environmental factors that need to be considered for a particular destination (Tosun, 2006; Scoones, 2009). The local community and local authorities have a better understanding of the positive and negative impacts of tourism in the destinations (Choi and Turk, 2011). It is essential to prioritize the development plans according to available resources. Directly or indirectly, the participatory approach paves the way for better strategies and policies for sustainable development in the destinations (Wood, 2002; Tsaur *et al.*, 2006). It generates opportunities to enhance economic benefits and socio-cultural inclusion through marketing locally produced goods and services. The local community, local tourism stakeholders and participating government agencies can develop a comprehensive vision about tourism development according to the needs and demands of the destination (Reed, 1997). It helps to build a common platform for presenting innovative ideas and suggestions, solving conflicts and discussing plans for the tourism industry. So, participatory design plays a vital role in developing a better conceptual framework through a critical review of every aspect of the destination (Timothy and Tosun, 2003).

In brief, the literature shows relatively limited examples of community consultation for participatory tourism development. To address this gap, in a developing context, this case study in Kerala, India explores the positive impacts of community participation on economic, socio-cultural and environmental factors through Responsible Tourism initiatives in Kumarakom destination. This research evaluates the effectiveness, fundamental elements and conceptual foundation of participatory design in the case study destination. The results of the case study indicate that participatory design can accelerate local community development, innovative initiatives, leadership, employment opportunities, demand for local products and sustainable development in the destination. The rest of the chapter is organized as follows. Section 1.1 provides a brief literature review and a description of the case study area and Responsible Tourism in Kerala. Section 1.2 discusses the methodology and methods used for data collection and analysis. Section 1.3 summarizes and discusses the case study findings. Lastly, the final section concludes the chapter.

1.1 Background

1.1.1 Review of literature

Destination design includes the strategic planning and organizing of hospitality services, tourist attractions, and cultural and experiential activities along with stakeholders in the destination (Hall and Page, 2003; Howie, 2003). Destination design is a complex framework as it considers every aspect of sustainable development in a destination (Fesenmaier and Xiang, 2017). Fry (2009) defines the designing concept as: 'it has infused every aspect of human conduct at the most basic level. – Our built environment, forms of work, modes of transport, manufactured products, media, infrastructure systems, and myriad other things are all designed concerning use. In actuality, the design is one of the main operative agents of the social, cultural, and economic functioning and dysfunction of humanity's made world' (p. 25). This holistic definition balances and evaluates various aspects and impacts of tourism from the primary level. In the context of urban tourism, it considers how the practical and logistical issues are managed by local authorities; for example, car parking, street cleaning, waste removal, control of crowds and visitors by the police during special events and environmental issues (Hall and Page, 2003; Howie, 2003; Fry, 2009). From a management perspective, it is defined as systematic planning and implementation of strategic plans in a tourist destination with management organizations, destination marketing, human resources and environmental management (Ritchie, 2004). In the destination, guests are linked with tourism products, services, physical boundaries, tourism stakeholders and host communities (Gnoth, 2002; Frischhut *et al.*, 2012).

Most of the growing literature and research focuses on expanding destination management concepts along with sustainable development in destinations (Komninos *et al.*, 2013). But the question is how to frame and conceptualize a smart framework; from whom we seek suggestions; by whom it should be implemented. Stakeholder theory is an essential component of effective destination design as it emphasizes the well-being of internal and external stakeholders in the destination (d'Angella and Go, 2009; Dwyer *et al.*, 2009). Participatory design is considered a new paradigm for integrating the activities and tourism stakeholders in the tourism destination (Boes *et al.*, 2015). The literature has defined three pillars of a smart destination, namely human capital, infrastructure and information (Komninos *et al.*, 2013). Hence, it is essential to include the human capital, namely the internal and external tourism stakeholders, in developing a framework in a destination. The participatory approach of tourism development is limited in the context of developing nations (Tosun, 2005). The significant complexities and difficulties to simultaneously relating tourism and local participation in developing nations are: lack of transparency, political instability, lack of information and undemocratic exceptional circumstances (Tosun, 2005, p. 337). The priorities of a smart destination can be evaluated by providing intelligent platforms to gather and distribute information among local stakeholders about tourists' travel experience (Nam and Pardo, 2011). The destination design assesses the knowledge about the demand and supply sides of available resources in the destination (Del Chiappa and Baggio, 2015). The destination design will include an effective and efficient solution for practical and logical issues in destinations (Hall and Page, 2003; Howie, 2003).

The direct involvement of local stakeholders and local government can accelerate economic development through local community empowerment and expanding achievable, sustainable tourism goals (Jackson and Morpeth, 1999). White (1982) argues that community participation is vital for tourism development as it helps to accomplish more results; services can be provided at lower cost; and intrinsic value for participants acts as a catalyst for growth, creates a sense of responsibility, ensures right ways to implement, guarantees the need, uses expertise and indigenous knowledge, and frees communities from dependence. Consequently, it helps to generate a sense of responsibility among the population and stimulates the implementation of better strategies in the tourism industry. Moreover, planning authorities can estimate and understand the local resources which are available for tourism development through consultation with local authorities and communities.

Participation is an instrument to educate and enlighten the local community and tourism stakeholders about rules and laws. Participatory design is heavily concentrated on developing an application framework or planning a practical design with the local community and tourism stakeholders, through reviewing the strengths, weakness, opportunities and threats (SWOT) of a tourism destination (Jackson and Morpeth, 1999). The participatory design should be incorporated with the destination design and management to achieve the competitive advantage of sustainable development in destinations.

1.1.2 Brief description of the case study area

India is a tourism hotspot blessed with tradition, cultural heritage, flora and fauna, and ancient monuments. The World Travel & Tourism Council (WTTC) estimated that between 2016 and 2026, India will top the list of the ten fastest-growing destinations for leisure travel, followed by Angola, Uganda, Brunei, Thailand, China, Myanmar, Oman, Mozambique and Vietnam. Kerala is one of the states in India at the forefront in the tourism and hospitality industry. Kerala enjoys unique geographic features and is known as 'God's Own Country' (Pulikottil, 2002; Vasudevan, 2008). Figure 1.1 highlights the location of Kerala State in India.

Kerala was formed as a state on 1 November 1956 from Travancore, Kochi and Malabar by combining Malayalam-speaking regions and now is divided into 14 districts (Ratcliffe, 1978; Franke and Chasin, 1992). Kerala experiences a humid, equatorial, tropical climate and lies at a latitude of 8°18′ to 12°48′N and a longitude of 74°52′ to 77°22′E (Kore *et al.*, 2016; Mathew and John, 2016). The total area of the state, which is about 38,864 km^2, is only 1.18% of the area of India. In 1986 the foreign tourist arrivals were 50,000, and Kerala reached a status of 1,096,407 international tourist arrivals in 2018 (Department of Tourism, 2018, p. 40). The total revenue from tourism during 2018 showed an increase of 8.61% over the previous year. The WTTC has selected Kerala as a partner state for enhancing the tourism industry (Chettiparamb and Kokkranikal, 2012). Kerala's unique selling proposition for marketing 'one of the ten paradises of the world' is ABC, an acronym for 'Ayurveda, Backwaters, Culture and cuisine' (Dhanesh, 2010).

Kumarakom is a destination comprising a cluster of islands that lie a few feet below sea level. The village covers an area of 51.67 km^2, which is inclusive of 24.13 km^2 of Vembanad Lake. Fishing, agriculture, shell mining and household industries are the main occupations. Records from the early 19th century indicate that the area was only sparsely populated. The 2001 census survey showed that the population had increased to about 23,000 and the number of households was approximately 5120 (Department of Tourism, 2007). Kumarakom has been declared a Special Tourism Zone by the Kerala State Government, as legislated for by the Kerala Tourism Act, 2005.

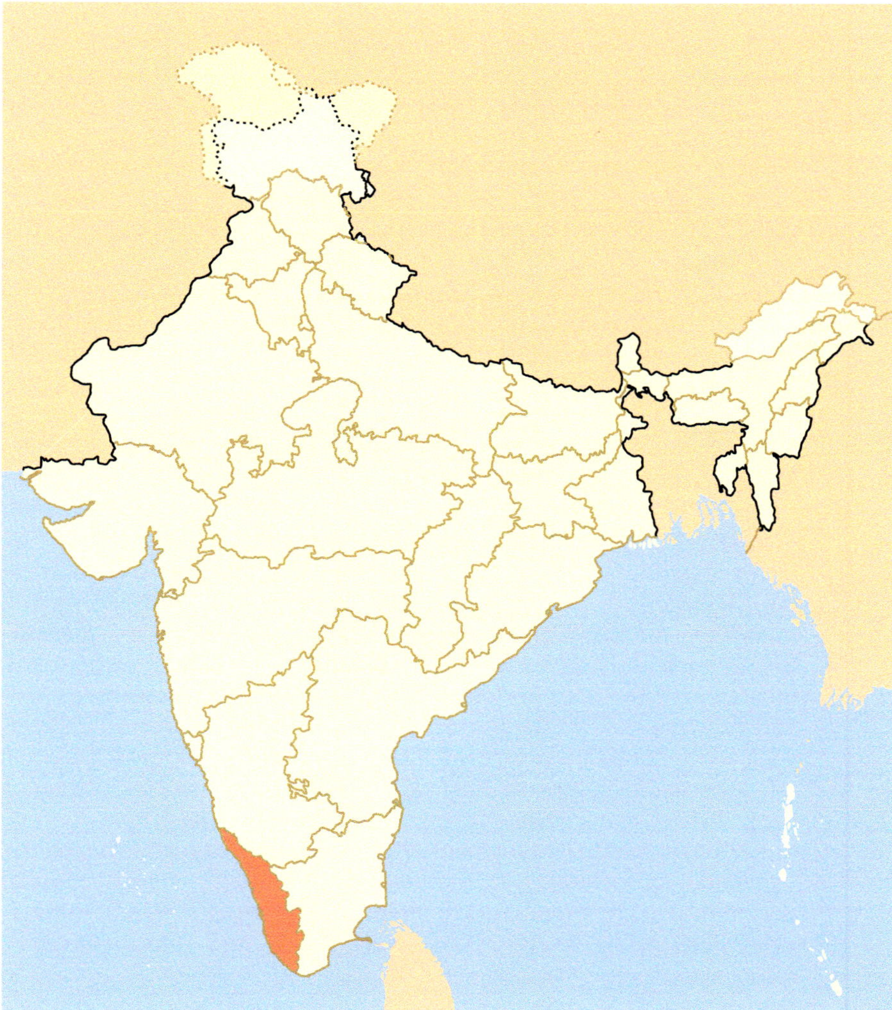

Fig. 1.1. Map showing location of Kerala (source: https://commons.wikimedia.org/wiki/File:India-KERALA.svg, accessed 4 November 2020).

1.1.3 Role of responsible tourism in Kerala

Responsible tourism is about making better places for people to live in and better places for people to visit (Cape Town Declaration, 2002). From a different point of view, responsible tourism is not a niche tourism product or brand but explains the ways responsible visitation can accelerate and foster a sense of responsibility in visitors and local people (Harrison and Husbands, 1996; Ebitu, 2010). It also focuses on three pillars of sustainable development, namely economic, social and environmental sustainability. Responsible tourism concentrates on fostering a sense of responsibility by visitors who enjoy the destination and enhancing the trust of local people (Goodwin and Francis, 2003; Frey and George, 2010). It focuses on inspiring visitors, local communities and stakeholders in the tourism industry to take more responsibility for making tourism sustainable. Responsible tourism is a local movement, being developed by different people in different ways around the world based on various local issues. It embraces all types of tourism and is focused on results (Leslie, 2012; Debicka and Oniszczuk-Jastrzabek, 2014). Researchers have studied responsible tourism investigating how it can enhance social sensitivity and ethical practices in tourism destinations (Goodwin and Francis, 2003; Frey and George, 2010; Debicka and Oniszczuk-Jastrzabek, 2014).

In 2007, responsible tourism initiatives were implemented by the tourism department in four destinations in Kerala, namely Kovalam, Kumarakom, Thekkady and Vythiri, as pilots. These responsible tourism initiatives contributed a better picture of the tourism industry with a participatory approach to the destination. Kumarakom destination won prestigious awards such as the United Nations World Tourism Organization (UNWTO) Ulysses Award for Innovation in Public Policy and Governance (2013), the National Award for Best Responsible Tourism Project (2007, 2009, 2013, 2014, 2018), the Pacific Asia Travel Association (PATA) Grand Award for Environment (2011), and the World Travel Market (WTM) Award for Best Poverty Reduction Project 2017 for the socio-economic impacts in the local community (UNWTO, 2017). UNWTO Regional Director for Asia and the Pacific, Xu Jing, congratulated Kerala for its successful implementation of the 'Responsible Tourism' model for other tourist destinations. The Director also noted: 'The projects can be included as part of UNWTO's global programmes and the report can be used as a case study for other destinations. We are eager to strengthen our relationship and discuss matters of common interest' (Ians, 2017). Kerala's Department of Tourism has as its primary aim to promote tourism to generate employment opportunities, reduce poverty and improve the livelihoods of local people. The government had taken initial steps in 2007 to constitute a State Level Responsible Tourism Committee (SLRTC) and Destination Level Responsible Tourism Committees (DLRTCs) to implement the concept of responsible tourism in destinations. The Kerala Institute of Tourism and Travel Studies (KITTS) was the nodal agency for responsible tourism initiatives to support local self-government and tourism stakeholders (Vijayakumar, 2013). The responsible tourism project was successful in generating employment opportunities and a better quality of life for the local community. In 2012, the initiative was extended to three more destinations, including Kumbalangi in Ernakulam, Vythiri and Ambalavayal in Wayanad, and Bekal in Kasaragod. In 2017, a new project called Participation for Planning and Empowerment through Responsible Tourism (PEPPER) was developed for planning and empowering the local people in participating in tourism development plans.

1.2 Research Methodology

The participatory approach and the destination design concept need to be analysed by exploring some of the thriving tourism initiatives in destinations.

This chapter is based on a case study of innovative initiatives and holistic insights of tourism stakeholders in a destination design. Responsible tourism cells are working in four tourism destinations in Kerala, and this case study was conducted in the destination Kumarakom. The success stories, awards and recognition of Kumarakom highlight the importance of cooperation and participation of stakeholders in destination management (Mathew and Kumar, 2014; Mathew and Sreejesh, 2017). The research adopted qualitative methods such as participant observation, interviews and focus groups for data collection. The data were collected in the period September to December 2017 by face-to-face interviews and focus group discussions. The primary method was in-depth semi-structured interviews, as this helps to understand and tries to explain a person's experience and his/her participation in destination design as a tourism stakeholder. The open-ended questions were focused on exploring the involvement of the local community in responsible tourism initiatives, the socio-economic benefits generation, addressing environmental sustainability issues and destination management through community participation.

Moreover, the research also focused on exploring the positive and negative impacts of participatory design in destination design and how it addressed the gaps and expectations of stakeholders in the destination. Twenty-seven semi-structured interviews were conducted with representatives of the 'Village Life Experience' package, hotels and restaurants, responsible tourism destination cells, local government authorities and local people in the destination. The interviewees were identified using a snowballing technique and by making initial contacts with tourism-related networks. Table 1.1 provides details about the number of interviewees included in each category. Additionally, a purposive sampling method was adopted for conducting interviews and focus group discussions with the local community and employees in boating, to obtain more in-depth data on their participation in the tourism industry. The research questions focused on understanding the contributions of the participatory approach, how it helped in destination design and how it helped to address the issues in the destination. Moreover, this provides an open space for a brainstorming discussion, helping to understand perspectives and insights about the concept among tourism stakeholders and the local community. All of the interviews and focus group discussions were recorded by tapes and notes and lasted between 45 and 90 minutes.

Table 1.1. Number of participants from each group (source: authors' own compilation).

Representation	No. of participants
Local government authorities	5
Responsible tourism destination coordinator and local community	7
'Village Life Experience' package	7
Safety and security officers in destination	3
Food and accommodation sector	3
Academic field	2
Total	27

The collected data were coded and analysed using NVivo 12 software. The analysis considers the interpretivist paradigm 'the world of human experience' and 'participants' views' (Mackenzie and Knipe, 2006) to recognize the impact of participatory design in destination management. The findings highlight the positive effects of responsible tourism initiatives on socio-economic aspects, environmental sustainability and destination management with participatory design. The majority of the participants emphasized that the tourism industry in Kumarakom had flourished with the introduction of responsible tourism and participatory design in the destination. The following section describes the significant findings and positive outcomes of the participatory approach in destination design in Kumarakom.

1.3 Findings of the Case Study

From interviews and focus group discussions, it is evident that in initial stages of tourism being introduced in Kumarakom, the local community welcomed the tourism industry with expectations of employment opportunities, income generation and appreciation of land value. But as tourism developed, the accommodation sector, including resorts and hotels, hired most employees from outside Kumarakom. As well, the clustering of resorts on the banks of backwaters was a significant hindrance to access by the local community for fishing, shell collection and other household activities (Mathew and Kumar, 2014). Hence the local community started to protest against the tourism industry with strikes and placards. In 2007, a responsible tourism initiative was commenced focused on generating positive impacts on economic, social and environmental sustainability. A DLRTC was developed to enhance the relationship between the industry and the local community to build trust (Venu and Goodwin, 2008; Mathew and Kumar, 2014).

The interviews with DLRTC officials, local community and focus group discussions provided the details about innovative activities, namely the 'Village Life Experience' package, homestays, vegetable farming, the vegetable-supplying unit Samruthi group, the Samruthi canteen and cultural shows in resorts in Kumarakom, which stimulated the potential advantages on rural development of involving local communities. Local authorities, tourism stakeholders and the women-oriented poverty reduction organization Kudumbashree took steps to implement responsible tourism initiatives (Mathew and Kumar, 2014). SWOT analysis with a participatory approach helped to reflect the local community perspective on tourism to integrate and maximize the positive impacts and minimize the adverse effects in the destinations. The focus group discussion with the local community helped to evaluate the effectiveness of participatory design in developing community initiatives with local people and local authorities. In the initial phase of developing the responsible tourism framework in 2007, 582 participants actively participated in the programme. The participants were from the local self-government (n = 45), people representatives (n = 10), partnering hotels (n = 135), Kudumbashree office bearers (n = 82), Samruthi activity group (n = 82), Kudumbashree micro-enterprises (n = 88) and Kudumbashree cultivation groups

($n = 140$) (Mathew and Kumar, 2014). The interviews and focus group discussions highlighted some of the positive contributions of participatory design, namely local community development and women's empowerment, innovative services and activities, local employment opportunities, entrepreneurship, demand for local products and sustainable development in the destination. These are discussed in the following sections.

1.3.1 Local community development and women's empowerment

The participants were not highly educated in Kumarakom, most of them based their livelihoods on fishing and agriculture. Interviews with DLRTC and SLRTC officials provided information about how they identified the obstacles and solutions for destination design. Negative attitudes towards entrepreneurship, unawareness of community benefits and insufficient capital for start-ups were the main obstacles in Kumarakom for the tourism industry. The tourism department, government and local authorities undertook initiatives to arrange language training, hospitality training classes and awareness programmes in Kumarakom to overcome hindrances. In the initial stages, the state government organized KITTS as the nodal agency of responsible tourism. KITTS arranged short-duration courses for local people who were interested in the tourism industry.

Kudumbashree was an active partner in implementation of the responsible tourism initiatives. Focus group discussions with the local community explained that around 1000 women are directly or indirectly linked with responsible tourism in Kumarakom. The responsible tourism initiatives were also focusing on encouraging vegetable farming and homestead farming and started a unit for producing vegetables and supplying them to hotel groups to generate income for the local community. This participatory approach helped to generate an optimal platform for creating demand for local products and services. Women started to engage in the supply of vegetables and performing the traditional art forms in hotels and resorts. Also, ten local women started a local ethnic restaurant with the name tag 'Samruthi Ethnic Food Restaurant' to provide local cuisine to visitors. The awareness programme and the active participation of local people in destination design helped to clarify their misconceptions about the tourism industry. Responsible tourism focuses on addressing economic, social and environmental issues and created responsibility among all tourism stakeholders in Kumarakom. Many participants suggested that the participatory approach made positive impacts on local community development and women's empowerment in Kumarakom:

> [the] local community understood responsible tourism is a better system. So, they started to engage in the tourism industry.
>
> (Participant 1 – government official)

> before this initiative, local people were against the tourism industry because they are not getting any benefit from it. With the introduction of responsible tourism,

> income [was] generated through creating demand for our products and opportunities for women to interact with the guests and explain about the culture, cuisine, and livelihoods such as coir making, weaving payya, etc.
>
> (Participant 2 – local community)

Government officials mentioned that the participatory approach played a significant role in women's empowerment and entrepreneurship development in Kumarakom. The local community and local authorities had arranged formal and informal meetings regularly to discuss demands, concerns and expectations of the local community and tourism stakeholders in Kumarakom. The participatory approach in Kumarakom helped address some of the significant interests and desires of tourism stakeholders and enhance local community development along with empowering women.

1.3.2 Leadership and community participation with innovative activities

The participatory approach helps to generate trust between community and tourism stakeholders to consider various factors of inclusive development in the destination. Kumarakom, the case study destination, is one of the best examples for showcasing innovative activities with the involvement of community participation. Leadership is a significant factor that contributed to the success of the responsible tourism framework in Kumarakom. Several group meetings were organized with local communities, local governments and tourism stakeholders for providing better understanding about the scope of tourism. The 'Village Life Experience' and 'A Day with Farmers' packages helped guests to better understand the culture and occupations of local people in the destination. It aided in marketing the rural destination by expanding the scope of local culture, enabling guests to experience farming and traditions, to generate economic benefits to the local community. These initiatives also paved the way for reducing poverty in the rural community.

The interviews and focus discussions revealed that local people took leadership to start micro-enterprises and be part of innovative initiatives. The responsible tourism framework in Kumarakom also integrated indigenous knowledge and traditional occupations of the local community, namely coir making, handicraft making, farming and fishing, as tourism products to offer guests. Interviews with DLRTC officials provided more clarity about the contribution of a participatory approach. More than 1000 families are directly or indirectly linked with responsible tourism in Kumarakom. Some of the activities in the 'Village Life Experience' package are coir making, coconut tree climbing, toddy tapping, fishing with bow and arrow, the weaving of payya, etc. 'A Day with Farmers' is a package for the guests to understand and experience authentic culture, farming and cultivation of crops. Guests are accompanied by an experienced local guide to provide information about the culture and traditions of Kumarakom. The amount collected from guests is equally distributed to the participants in these two packages. The participation of local people in the tourism industry helps to increase their leadership and entrepreneurship skills, providing additional income and better quality of life. These innovative activities paved the way for better destination design by linking the tourism industry and the local community. Nearly 2317 participants actively engaged in responsible tourism initiatives (Mathew and Kumar, 2014).

The majority of participants suggested responsible tourism was the first project which took the initiative to improve leadership skills among community members as well as their participation in the tourism industry in Kumarakom:

> strengthen our leadership skills through training and awareness classes, supported start-up units.
>
> (Participant 4 – focus group discussion – local community)

> they actively participated in innovative initiatives such as 'Village Life Experience' and 'A Day with Farmers'. It helps to generate employment opportunities, poverty alleviation and income by demonstrating their occupation to guests.
>
> (Participant 5 – government official)

The findings from the interviews and focus group discussions clearly illustrate that responsible tourism initiatives opened innovative and collaborative projects with the local community to provide a better experience to guests through two packages, 'Village Life Experience' and 'A Day with Farmers'. Besides, interviews with local government officials mentioned the participation of the local community as an opportunity to explore the potential economic, socio-cultural and environmental advantages of enhancing innovative activities and tourism products in Kumarakom.

1.3.3 Employment generation for local people

The local community participation also paved the way for employment generation in the destination. DLRTCs arranged classes on candle making, handicraft making, cooking, cultural arts and paper bag making, as well as language training programmes to develop skills and confidence to interact with visitors. The hospitality industries started to hire local people as employees. The responsible tourism project identified around 25 local guides to explain the peculiarities and culture of the destination to guests. The responsible tourism project undertook initiatives to promote employment opportunities to the local community in homestays, houseboats, vegetable farming, etc. A directory was introduced with responsible tourism, which provides personnel details of skilled, unskilled and semi-skilled employees in Kumarakom. This innovative idea provides more job opportunities for the local community. The tourism industry partners can refer to the directory and make arrangements with the local community when there is a need for skilled and unskilled jobs. Moreover, to understand the resource availability, the responsible tourism cell conducted resource mapping in Kumarakom *panchayat*. This helped to record the data related to the culture, arts, human resources, profiles, etc. Some of the participants emphasized that it paved a new outlook to tourism stakeholders to include local skilled, unskilled and semi-skilled employees in the tourism industry:

providing an opportunity for skilled, unskilled, and semi-skilled employees to link with the tourism stakeholders in Kumarakom. E.g. more income for local taxi drivers, plumbers, electricians in Kumarakom.

(Participant 6 – government official)

1.3.4 Better strategic plans to create demand for local products in the tourism industry

The participatory approach helped to generate an optimal platform for creating demand for local products and services in Kumarakom. The production and procurement of vegetables and other local products in Kumarakom are managed by a separate group called Samruthi (Mathew and Kumar, 2014). From the interviews and focus group discussions, it is evident that the participatory design encouraged the local community in agricultural farming and production of small-scale products in Kumarakom. Interviews with DLRTCs and small units highlighted a production procurement system called 'Samruthi', which was developed with different committees to control prices and assure the quality of the local products. The Samruthi group acts as an intermediary between the hotel industry and local farmers in Kumarakom in the supply of vegetables, candles, paper bags, etc. Figure 1.2 illustrates the role of the Samruthi group. Responses from participants indicate that Samruthi took the initiative to reduce the gap between the local production and supply of local products in Kumarakom:

encourage the resorts to use locally available products such as papaya, coconut, banana leaves, candles, paper bags, etc., to support local community.

(Participant 7 – government official)

it helps to create demand for our products and encourage us to start agricultural farming. Local government authorities arranged classes on farming, cultivation, etc.

(Participant 8 – local community)

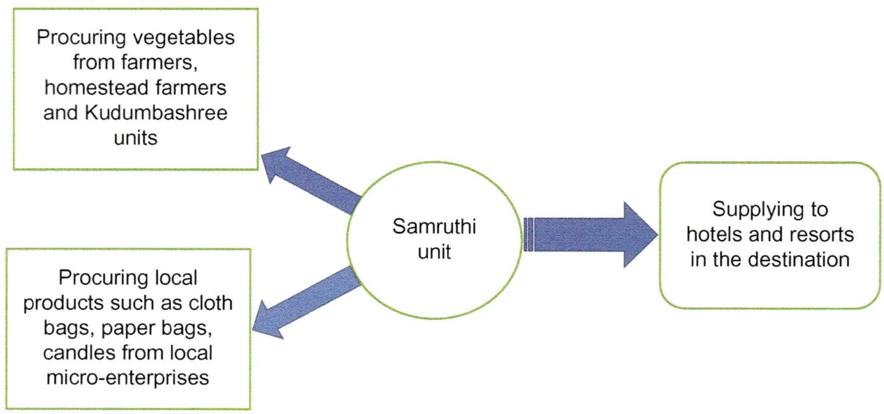

Fig. 1.2. The role of the Samruthi group (source: authors' own elaboration).

Interviews with SLRTC and DLRTC officials explained that the significant concerns during the initial stages of the Samruthi group were on assuring product quality and fixing prices for local products. The participatory approach resolved the problem by developing a price fixation committee and a quality assurance committee to check the quality of products. The price fixation committee is constituted by the responsible tourism cell, the department of tourism, Kudumbashree members, hotel and resort representatives, and the *grama panchayat*. The quality assurance committee represents hotel chefs, tourism cells, the department of tourism, Kudumbashree members, the health inspector, agricultural officer and the *grama panchayat*. After identifying the significant contribution of a participatory approach, a new project called PEPPER has been introduced to enhance destination design from the grassroot level.

1.3.5 Encouraging sustainable practices

The concept of participatory design has contributed to some innovative practices in sustainable development in Kumarakom. From the focus group discussions, it was clear that the local community and the environmentalist analysed the negative impacts of the tourism industry in Kumarakom. The responsible tourism framework addressed some sustainability issues in Kumarakom through awareness programmes and campaigns to conserve natural habitats, control water pollution and regenerate mangrove trees. With the clustering of resorts and hotels, the number of mangrove trees in Kumarakom was drastically reduced, which in turn adversely affected natural habitats, the breeding of fish and soil erosion. In the interviews and focus group discussions, the participants highlighted that this was identified by the local community in Kumarakom. To overcome the adverse impacts, responsible tourism introduced a 'Mangrove regeneration campaign' in Kumarakom to create awareness among tourism stakeholders to conserve mangroves. The local self-government also supported responsible tourism initiatives, and around 1500 mangroves have been planted in Kumarakom. The bird sanctuary at Kumarakom was declared a plastic-free zone to control plastic waste in natural habitats for migrating birds. Many participants suggested that the participatory approach contributed to some solutions and better practices for sustainable development in Kumarakom:

> issues such as water pollution, reduction in the fish population, soil erosion, etc., as most of the local community live their livelihoods on fishing and agriculture. So, it was easy for them to identify problems.
>
> (Participant 9 – government official)

> took the initiative to conserve the resources in Kumarakom through campaigns and awareness programs, e.g. authorities will charge fines to guests who drop plastic waste and cans in [the] bird sanctuary.
>
> (Participant 10 – local community)

Responsible tourism promotes country boat rides to control water pollution and to generate local employment opportunities in Kumarakom. On the other hand, the majority of the respondents highlighted that water bodies became polluted with diesel, oils, food waste and plastic waste, which affected fish biodiversity, with the expansion of the tourism business.

To summarize, the above findings from interviews and focus group discussions clearly illustrate that a participatory approach can contribute to better innovative tourism products and ideas for destination design. Along with local authorities, tourism stakeholders and local communities developed a framework that helps to enrich better visitor experiences and destination design. Moreover, the case study has assessed perceptions of the local community in the tourism industry. Most of the respondents explained that this project was successful in Kumarakom as it considered the participation of all the tourism stakeholders in the destination. The case study has also attempted to examine the perceived equitable distribution of benefits of the participatory approach in destination design. It identified the positive impacts, expectations, community engagement and strategies adopted for a better destination design with the participation of tourism stakeholders in Kumarakom.

Discussion and Conclusions

This chapter helps explore a better understanding of the scope of community participation in tourism planning with a case study in Kerala. From the findings, it is evident that the introduction of a practical model of responsible tourism was excellent, as the local community became a part of the tourism industry. Tourism is one of the primary industries which can directly generate employment opportunities and enhance better livelihoods for residents, therefore being considered a 'passport to development', so tourism projects need to be planned with more participatory approaches and attention (Liu and Wall, 2006, p. 169). Some of the significant case study findings indicate that this initiative has built confidence among the local community to engage in the tourism industry. Besides, the responsible tourism initiatives created demand for local products, better livelihoods and collaborative projects. These initiatives focused on generating positive economic impacts on the local community and building trust and group effort with tourism stakeholders. Directly or indirectly, these initiatives helped to develop strong links between the tourism industry and local people. With the responsible tourism cell intervention, local agricultural products were identified that can be supplied regularly to the hotel industry. The majority of respondents from the 'Village Life Experience' package and local community in Kumarakom mentioned that responsible tourism initiatives helped the local community to earn an additional income. Some responsible tourism strategies promoted the participation of women and minority groups through awareness and skill development programmes. However, local communities and government officials highlighted that the tourism industry contributes vast waste and water pollution in Kumarakom. The above findings and quotes illustrate that DLRTCs in Kumarakom took some steps to support biodiversity by arranging campaigns, raising awareness and encouraging eco-friendly practices. On the other hand, the respondents also highlighted some areas that need further attention, such as water pollution, plastic waste, pesticides, insecticides and fish scarcity, considered as significant threats to biodiversity, to control the deterioration of the quality of the environment.

Destination design is essential for tourism growth and the sustainable development of a destination. Human capital, leadership, social capital and innovation are considered the four fundamental factors for developing a smart tourism destination (Boes *et al.*, 2015). Community participation is regarded as a better strategy for identifying underlying factors, demands and expectations of tourism stakeholders and the local community in a destination. It helps to generate a common platform for the tourism industry, tourism and other related government authorities and the local community to develop a better framework for sustainable development in a destination. Participation creates responsibility, understanding of indigenous knowledge, socio-cultural inclusion, better strategic plans and sustainable development among tourism stakeholders in the destination. In this chapter, we examined the significant contribution of the participatory design of responsible tourism in Kumarakom destination. This case study also addressed how to overcome those issues such as unemployment and poverty, and how to interlink initiatives with the local community in the destination. The participants in the case study highlighted that the tourism industry in Kumarakom had flourished with the introduction of community initiatives. Participation and integration of the local community paved the way for innovative initiatives, unique tourism products and shaped positive attitudes towards sustainable development in the tourism industry. To conclude, the case study found that participatory design contributed to local community development, employment generation, women's empowerment, demand for local products and innovative practices on sustainable development in Kumarakom destination.

It is also acknowledged that several limitations bind the case study. The critical limitations are related to the generalizability of the findings and possible sample bias. A qualitative approach was used for addressing the research aim, which has limitations on the overall generalizability of findings. So, future research should consider quantitative methods for higher levels of reliability and low bias of data. However, this case study certainly revealed many future trends and some significant findings which indicate we can extend the scope of academic research towards policies and environmental sustainability for framing better strategies. Through analysing natural resource management of tourism initiatives, we can help identify better strategies for mitigating environmental sustainability challenges and for adaptation actions.

References

Boes, K., Buhalis, D. and Inversini, A. (2015) Conceptualising smart tourism destination dimensions. In: Tussyadiah, I. and Inversini, A. (eds) *Information and Communication Technologies in Tourism 2015: Proceedings of the International Conference in Lugano, Switzerland, February 3–6, 2015*. Springer, Cham, Switzerland, pp. 391–403.

Cape Town Declaration (2002) Cape Town declaration on responsible tourism. Cape Town Conference on Responsible Tourism in Destinations. Available at: https://responsibletourismpartnership.org/cape-town-declaration-on-responsible-tourism/ (accessed 16 November 2020).

Chettiparamb, A. and Kokkranikal, J. (2012) Responsible tourism and sustainability: the case of Kumarakom in Kerala, India. *Journal of Policy Research in Tourism, Leisure and Events* 4(3), 302–326.

Choi, H.C. and Turk, E.S. (2011) Sustainability indicators for managing community tourism. In: Budruk, M. and Phillips, R. (eds) *Quality-of-Life Community Indicators for Parks, Recreation and Tourism Management*. Social Indicators Research Series, vol. 43. Springer, Dordrecht, the Netherlands, pp. 115–140.

Clark, G., Kosoris, J., Hong, L.N. and Crul, M. (2009) Design for sustainability: current trends in sustainable product design and development. *Sustainability* 1(3), 409–424.

Crouch, G.I. and Ritchie, J.B. (1999) Tourism, competitiveness, and societal prosperity. *Journal of Business Research* 44(3), 137–152.

D'Angella, F. and Go, F.M. (2009) Tale of two cities' collaborative tourism marketing: towards a theory of destination stakeholder assessment. *Tourism Management* 30(3), 429–440.

Debicka, O. and Oniszczuk-Jastrzabek, A. (2014) Responsible tourism in Poland. In: *22nd Biennial International Congress, Tourism and Hospitality Industry 2014 (THI2014), Trends and Challenges, Opatija, Croatia, 8–9 May 2014*, pp. 189–202. Available at: https://thi.fthm.hr/congress-proceedings/category/2-2014 (accessed 5 November 2020).

Del Chiappa, G. and Baggio, R. (2015) Knowledge transfer in smart tourism destinations: analyzing the effects of a network structure. *Journal of Destination Marketing & Management* 4(3), 145–150.

Department of Tourism (2007) *Kumarakom Geographic and Demographic Details*. Government of Kerala, Kerala, India.

Department of Tourism (2018) *Kerala Tourism Statistics 2018*. Available at: https://www.keralatourism.org/tourismstatistics/tourist_statistics_2018_book20191211065455.pdf (accessed 22 March 2019).

Dhanesh, G.S. (2010) Kerala, God's Own Country, from backwaters to one of the ten paradises on earth. In: Moss, D., Powell, M. and DeSanto, B. (eds) *Public Relations Cases: International Perspectives*. Routledge, Abingdon, UK, pp. 40–58.

Dredge, D. (1999) Destination place planning and design. *Annals of Tourism Research* 26(4), 772–791.

Dwyer, L., Edwards, D., Mistilis, N., Roman, C. and Scott, N. (2009) Destination and enterprise management for a tourism future. *Tourism Management* 30(1), 63–74.

Ebitu, E. (2010) Promoting an emerging tourism destination. *Global Journal of Management and Business Research* 10(1), 21–28.

Fesenmaier, D.R. and Xiang, Z. (2017) Introduction to tourism design and design science in tourism. In: Fesenmaier, D. and Xiang, Z. (eds) *Design Science in Tourism: Foundations of Destination Management*. Tourism on the Verge Series. Springer, Cham, Switzerland, pp. 3–16.

Franke, R.W. and Chasin, B.H. (1992) Kerala State, India: radical reform as development. *International Journal of Health Services* 22(1), 139–156.

Frey, N. and George, R. (2010) Responsible tourism management: the missing link between business owners' attitudes and behaviour in the Cape Town tourism industry. *Tourism Management* 31(5), 621–628.

Frischhut, B., Stickdorn, M. and Zehrer, A. (2012) Mobile ethnography as a new research tool for customer-driven destination management – a case study of applied service design in St. Anton, Austria. In: *CAUTHE 2012: The New Golden Age of Tourism and Hospitality; Book 2; Proceedings of the 22nd Annual Conference*. La Trobe University, Melbourne, Australia, pp. 161–167.

Fry, T. (2009) *Design Futuring: Sustainability, Ethics and New Practice*. University of New South Wales Press, Sydney, Australia, pp. 71–77.

Gnoth, J. (2002) Leveraging export brands through a tourism destination brand. *Journal of Brand Management* 9(4), 262–280.

Goodwin, H. and Francis, J. (2003) Ethical and responsible tourism: consumer trends in the UK. *Journal of Vacation Marketing* 9(3), 271–284.
Hall, C.M. and Page, S. (2003) *Managing Urban Tourism*. Prentice-Hall, Harlow, UK.
Harrison, L.C. and Husbands, W. (1996) *Practicing Responsible Tourism: International Case Studies in Tourism Planning, Policy and Development*. Wiley, New York.
Howie, F. (2003) *Managing the Tourist Destination*. Continuum, London.
Ians, J. (2017) *Kerala's 'Responsible Tourism' initiative a role model, says UNWTO. India Today*, 4 February 2017. Available at: https://www.indiatoday.in/travel/india/story/kerala-responsible-tourism-role-model-unwto-asia-india-lifetr-958972-2017-02-04 (accessed 5 November 2020).
Jackson, G. and Morpeth, N. (1999) Local Agenda 21 and community participation in tourism policy and planning: future or fallacy. *Current Issues in Tourism* 2(1), 1–38.
Komninos, N., Pallot, M. and Schaffers, H. (2013) Special issue on smart cities and the future internet in Europe. *Journal of the Knowledge Economy* 4(2), 119–134.
Kore, S.B., Lole, A., Gunjagi, D.A. and Shinde, S.S. (2016) Feasibility of offshore wind farm in India. *International Research Journal of Engineering and Technology (IRJET)* 7(11), 995–998.
Leslie, D. (2012) *Responsible Tourism: Concepts, Theory and Practice*. CAB International, Wallingford, UK.
Liu, A. and Wall, G. (2006) Planning tourism employment: a developing country perspective. *Tourism Management* 27(1), 159–170.
Mackenzie, N. and Knipe, S. (2006) Research dilemmas: paradigms, methods and methodology. *Issues in Educational Research* 16(2), 193–205.
Mathew, L.R. and John, D. (2016) Frugal automation of sustainable practices in Kerala. *Procedia Technology* 24, 1211–1218.
Mathew, P. and Kumar, R. (2014) Responsible tourism – a grass root level empowerment mechanism: case study from Kerala. *Innovative Issues and Approaches in Social Sciences* 7(1), 53–73.
Mathew, P.V. and Sreejesh, S. (2017) Impact of responsible tourism on destination sustainability and quality of life of community in tourism destinations. *Journal of Hospitality and Tourism Management* 31, 83–89.
Matsumoto, M., Umeda, Y., Masui, K., Fukushige, S. Umeda, Y. (2012) *Design for Innovative Value Towards a Sustainable Society. Proceedings of EcoDesign 2011: 7th International Symposium on Environmentally Conscious Design and Inverse Manufacturing*. Springer, Dordrecht, the Netherlands.
Nam, T. and Pardo, T.A. (2011) Conceptualizing smart city with dimensions of technology, people, and institutions. In: *Proceedings of the 12th Annual International Digital Government Research Conference: Digital Government Innovation in Challenging Times, College Park, Maryland, USA, June 2011*, pp. 282–291. Available at: https://dl.acm.org/doi/10.1145/2037556.2037602 (accessed 5 November 2020).
Paget, E., Dimanche, F. and Mounet, J.-P. (2010) A tourism innovation case: an actor–network approach. *Annals of Tourism Research* 37(3), 828–847.
Pulikottil, P. (2002) As East and West met in God's Own Country: encounter of western pentecostalism with native pentecostalism in Kerala. *Asian Journal of Pentecostal Studies* 5(1), 5–22.
Ratcliffe, J. (1978) Social justice and the demographic transition: lessons from India's Kerala State. *International Journal of Health Services* 8(1), 123–144.
Reed, M.G. (1997) Power relations and community-based tourism planning. *Annals of Tourism Research* 24(3), 566–591.
Ritchie, B.W. (2004) Chaos, crises and disasters: a strategic approach to crisis management in the tourism industry. *Tourism Management* 25(6), 669–683.
Schaffers, H., Komninos, N., Pallot, M., Trousse, B., Nilsson, M. and Oliveira, A. (2011) Smart cities and the future internet: towards cooperation frameworks for open innovation. In: Domingue, J.,

Galis, A., Gavras, A., Zahariadis, T., Lambert, D., Cleary, F., *et al.* (eds) *The Future Internet. Future Internet Assembly 2011: Achievements and Technological Promises*. Lecture Notes in Computer Science, vol. 6656. Springer, Berlin/Heidelberg, Germany, pp. 431–446.

Scoones, I. (2009) Livelihoods perspectives and rural development. *Journal of Peasant Studies* 36(1), 171–196.

Timothy, D.J. and Tosun, C. (2003) Arguments for community participation in the tourism development process. *Journal of Tourism Studies* 14(2), 2–15.

Tosun, C. (2005) Stages in the emergence of a participatory tourism development approach in the developing world. *Geoforum* 36(3), 333–352.

Tosun, C. (2006) Expected nature of community participation in tourism development. *Tourism Management* 27(3), 493–504.

Tsaur, S.H., Lin, Y.C. and Lin, J.H. (2006) Evaluating ecotourism sustainability from the integrated perspective of resource, community and tourism. *Tourism Management* 27(4), 640–653.

UNWTO (2017) *UNWTO Annual Report 2016*. United Nations World Tourism Organization, Madrid.

Vasudevan, S. (2008) The role of internal stakeholders in destination branding: observations from Kerala Tourism. *Place Branding and Public Diplomacy* 4(4), 331–335.

Venu, V. and Goodwin, H. (2008) The Kerala Declaration on Responsible Tourism. In: *Incredible India! Second International Conference on Responsible Tourism in Destinations, Kerala, 21–24 March 2008*. Available at: http://www.haroldgoodwin.info/documents/KeralaDeclaration.pdf (accessed 5 November 2020).

Vijayakumar, S.R.B.R. (2013) *Tourism and Livelihood*. Kerala Institute of Tourism and Travel Studies (KITTS), Thiruvananthapuram, India.

White, A.T. (1982) Why community participation? A discussion of the arguments. *Assignment Children* 59–60, 17–34.

Wood, M. (2002) *Ecotourism: Principles, Practices and Policies for Sustainability*. United Nations Environment Programme, Paris.

2 Planning Agritourism Development in Sri Lanka: Farmers' Incentives and Barriers

Frank Niranjan[1] and Glen Croy[2*]

[1]*Ministry of Agriculture, Sri Lanka;* [2]*Monash University, Australia*

Introduction

Rural areas are facing growing pressures including population loss, small-scale settlements and susceptibility to great fluctuations in commodity prices (Lane, 1994; Liu, 2006; Ollenburg and Buckley, 2007). Generally, due to these pressures, rural areas have limited opportunities for economic development; crops and livestock are the traditional and primary sources of income (Ollenburg and Buckley, 2007). To stimulate development, it is important for planning and development officials and farmers to examine alternative uses of rural resources (Koscak, 1998). These same indicative pressures, including low population density, open space, natural environment and small-scale settlements, are also seen as beneficial for tourism opportunities (Lane, 1994; Liu, 2006). As such, in response to these rural challenges, planning and development officials have instigated a number of viable development alternatives, including tourism (Ollenburg and Buckley, 2007; Gao *et al.*, 2014; Barbieri *et al.*, 2016). None the less, tourism is a sector reliant on different skills from traditional agriculture (Fleischer and Felsenstein, 2000). This causes perceptual barriers to agritourism and the capacity to exploit the new sector, especially by those who control the resources – the farmers. Additionally, agritourism employment opportunities are generally seasonal and with relatively low wages (Fredrick, 1993). Furthermore, although agritourism is described as an environmentally friendly business, it can reduce access to valuable and finite resources in rural areas (Gibson, 1993), and while agritourism creates new demand, it displaces existing demand in the rural economy (Hoy, 1996). Hence, a more detailed

*E-mail: glen.croy@monash.edu

DOI: 10.1079/9781789246698.0002

review of the resource, agriculture, environment and cultural base is required before selecting agritourism as a development response to rural area challenges (Liu, 2006). Moreover, while studies have been undertaken to find out residents' attitudes towards tourism development (Choo and Petrick, 2014), few studies have attempted to specifically understand the farmer's perspective of agritourism (Kunasekaran *et al.*, 2011; Tew and Barbieri, 2012).

By better understanding farmers' perceptions of agritourism, planning and development, officials are better able to make appropriate and facilitating decisions (Ollenburg and Buckley, 2007). These challenges are more prominent in rural areas of emerging economies, such as South Asia, where institutional change and liberalization have magnified the aforementioned pressures. However, the pressures have also introduced emergent tourism demand for rural areas (Kunasekaran *et al.*, 2011; Malkanthi and Routry, 2011; Tiraieyari and Hamzah, 2012). Hence, we aim to understand farmers' perceptions of agritourism in emerging economies of South Asia, using Sri Lanka as a focus case. In achieving the aim, we anticipate contributing to understanding agritourism within emerging economies in three ways. First, we propose to identify farmers' propensity for agritourism as an option (Tew and Barbieri, 2012; Barbieri, 2013). Second, we will identify the most prominent factors motivating farmers to engage in agritourism (Ollenburg and Buckley, 2007; Tew and Barbieri, 2012). Finally, we will highlight priority agritourism themes for planning and development officials in an emerging economy context (Ollenburg and Buckley, 2007).

To achieve this aim, we review the literature on agritourism, farm and farmers, resources, and incentives and barriers for agritourism, followed by discussing Sri Lankan agritourism. We then outline the method used in this research and subsequently discuss the results. We present an agritourism development understanding and the implications, and finally conclude the chapter.

2.1 Literature Review

2.1.1 Agritourism

When the financial situation of a farm is constrained, there are pressures to look outside agriculture to maintain the farmer's (and family's) operations and lifestyle. One option is agritourism (Ollenburg and Buckley, 2007; Barbieri *et al.*, 2016). Indeed, agritourism is implemented as a development tool in many parts of the world (Barbieri, 2013; Barbieri *et al.*, 2016), including in Italy, New Zealand, the USA and Scotland, as it improves farm revenue and assists farmers' economic situation (Fleischer and Felsenstein, 2000). Importantly, agritourism also enhances the quality of life for the farm family and wider community in non-economic terms (Ollenburg and Buckley, 2007; Barbieri, 2013; Gao *et al.*, 2014). Indeed, a range of agritourism purposes and outcomes are identified (Table 2.1).

However, agritourism is discussed using a variety of interchangeable terms, including agritainment, farm-based tourism, agrotourism, agricultural tourism, farm visits and agritourism (Barbieri *et al.*, 2016). After prompting, potential

Table 2.1. Agritourism purposes and outcomes for farmers, rural communities and tourists (source: authors' own compilation).

Farmers	Rural communities	Tourists
Economic benefits	Enhances rural economic development	
Creating employment opportunities for family	Retains the next generation in the rural areas instead of migrating to urban areas	
Preserving tangible heritage in their farmlands	Attracts urban dwellers into rural areas, enhancing the sense of place	Provides city dwellers an opportunity to escape from urban environments to discover their rural roots
Educating guests and showing the farm		Educational and recreational benefits
Initiating commercial channels for local produce		Access to fresh local food items
Converting unproductive land into areas of opportunity		
Diversification of farm activities		
Meeting people		

tourists preferred 'farm visit' (and 'agritourism' ranked last). However, and interestingly, agritourism was the term that most resonated with farmers (Barbieri *et al.*, 2016) and is also the term adopted in this chapter.

Preferred agritourism definitional characteristics include agricultural setting, entertainment, farm and education (Lane, 1994; Phillip *et al.*, 2010; Gil Arroyo *et al.*, 2013; Barbieri *et al.*, 2016; Kline *et al.*, 2016). Gil Arroyo *et al.* (2013) also find that residents, farmers and public officers agree agritourism included authentic activities carried out on working farms and reject activities offered on non-working farms (also see Lane, 1994; Nickerson *et al.*, 2001; Phillip *et al.*, 2010; Tew and Barbieri, 2012; Gao *et al.*, 2014). Drawing from these studies, agritourism is defined as tourist interaction with agriculture on working farms.

2.1.2 Agritourism farms, farmers and resources

Previous studies demonstrate that certain farms and farmers are more likely to undertake agritourism, indicating that tourism is not a universal solution to the pressures rural areas are facing (Lane, 1994). Studies demonstrate that agritourism is more likely on smaller farms (Loureiro and Jervell-Moxnes, 2004; Ollenburg and Buckley, 2007; Bagi and Reeder, 2012; Joo *et al.*, 2013; Gao *et al.*, 2014), close to urban areas (Loureiro and Jervell-Moxnes, 2004; Bagi and Reeder, 2012), that are accessible (Kunasekaran *et al.*, 2011; Bagi and Reeder, 2012), involved in wider networks (such as conservation) (Lane, 1994; Mnguni, 2010; Bagi and Reeder, 2012) and in tourism areas (Kunasekaran *et al.*, 2011).

Bagi and Reeder (2012) further find that farms paying for farm advice or that are partnership or corporate farms were more likely to participate in agritourism. The small size of the farms participating in agritourism may indicate greater exposure to the previously noted pressures, combined with the limited agricultural opportunities to respond to them (Ollenburg and Buckley, 2007; Joo *et al.*, 2013; Kline *et al.*, 2016).

Further, a small number of studies investigate agritourism farmers' characteristics, emphasizing certain demographics. Agritourism farmers have been found to be younger (Loureiro and Jervell-Moxnes, 2004) and have higher education levels (Bagi and Reeder, 2012; Mnguni, 2010; Joo *et al.*, 2013). Additionally, Loureiro and Jervell-Moxnes (2004) further find an increased participation in agritourism if there was a female presence on the farm, although Joo *et al.* (2013) find this participation to be from older females.

The characteristics of agritourism farms and farmers provide an indication for potential compatibility, interest and success, although, even with interest, the opportunity might well be limited by resources needed and available. López and García (2006) broadly highlight a rural environment, agricultural equipment and accommodation as the resources needed for agritourism (also see Gao *et al.*, 2014). More specifically, resources are identified by what agritourists do, largely from North American studies (Table 2.2).

The range of agritourism activities indicates a specifically rural experience, contrasted with those available in urban environments, making the landscape context crucial (Gao *et al.*, 2014). The activities themselves are then the means of authentic agricultural interaction. The resources indicate low levels of complexity; however, and importantly, for agritourism to be appropriately resourced, there is a need for people and capacity on the farms to supply the experiences. Thus, it is especially important to investigate the incentives and barriers for farmers to participate.

2.1.3 Agritourism incentives and barriers to participate

Incentives and barriers to farmers' participation in agritourism are important and are described in previous studies. Hall and Jenkins (1997) outline that the government policy environment can be very influential to incentivize agritourism establishment. They highlight regulation, voluntary instruments, expenditure, financial incentives and non-intervention decisions all as means

Table 2.2. Agritourism tourist activities (source: authors' own compilation).

Casual photography	Children's educational programmes	Christmas tree sales	Farm tours
Hayrides	Horseback riding	Hunting and fishing	On-farm festivals and events
Overnight stays on farms	Participating in rituals	Petting zoos	Pick-your-own fruit and vegetables
Staying in tree houses	Swimming	Viewing wildlife	Watching fish and birds

to engage regions and farmers in agritourism. None the less, the specific policy direction and content will be determined by the context and what incentivizes farmers directly (Fleischer and Felenstein, 2000).

Enhanced farm profitability is one of the important incentives for the adoption of agritourism (Joo *et al.*, 2013; Kline *et al.*, 2016). However, while the resource profile is not complex, it likely will still require some investment, which may present a barrier to establish and maintain an agritourism venture (Liu, 2006). Furthermore, agritourism contributes relatively little extra to farm incomes (Hjalager, 1996). Fleischer and Pizam (1997) propose that this might be due to the small-scale nature of agritourism and supplying products and services to a highly seasonal market. None the less, studies show that profit impacts are highest among small-scale farms, generating the greatest household incomes and returns to assets (Joo *et al.*, 2013; Schilling *et al.*, 2014), and the impact for larger farms is not statistically significant (Schilling *et al.*, 2014). Despite the impact effects, Ollenburg and Buckley (2007) find that those on larger farms are more motivated by supplementing income than those on smaller farms.

Another incentive is that agritourism facilitates the diversification and even expansion of the rural economy (Sharpley, 2002; Ollenburg and Buckley, 2007; University of California Small Farm Program (UCSFP), 2011; Barbieri *et al.*, 2016). The diversification may mean less dependence on traditional challenged markets, protection of traditions and crafts, development of new markets and new employment opportunities for operators and family members (Carter, 1998; Ollenburg and Buckley, 2007; Gao *et al.*, 2014; Kline *et al.*, 2016). Sharpley (2002) emphasizes that agritourism can also be a means to protect and improve the agricultural environment and infrastructure (also see Kline *et al.*, 2016 regarding sustainable farming). At times agritourism activities are created with little extra effort to earn more money for the operator (Bagi and Reeder, 2012). However, for most it is a transitional process that requires building tourism and business skills and capacities.

On the tourism skills side, the combined supply of agricultural commodities with tourist-service demands may be incompatible and agritourism businesses may find it difficult to execute the service role (Fleischer and Pizam, 1997). Furthermore, the supplied products and services may not match tourists' expectations (Roberts, 1996) and agritourism can only be competitive if tourists value it (Fratu and Gruia, 2012). Rural areas are not equally attractive and unless tourists are provided with the total package, it is difficult to attract them to rural areas (Gannon, 1994).

Service dissonance is mainly due to farmer unfamiliarity with tourism activities, difficulty in supplying skilled labour, insufficient attention to local differences in capacity building, and when training courses are offered they may not address the local skill-gap needs and challenges (Liu, 2006; Kunasekaran *et al.*, 2011). Sharpley (2002) further highlights that rural areas are additionally limited by the amount and access to support and training. Reflecting employment opportunities, agritourism is also criticized for the creation of low-paying and seasonal work, potentially creating an underclass through underemployment (Deller, 2010). Countering this is the potential of family supplementary income through agritourism, as part of the diversified farm profile, providing the opportunity to undertake the main cultivation operation successfully.

On the business skills side, Embacher (1994) notes that generally agritourism enterprises do not develop sufficient skills or allocate sufficient resources for effective marketing (also see Sharpley, 2002). This is further complicated as Tew and Barbieri (2012) suggest that an incentive for starting agritourism is as a marketing tool for agricultural produce. To enhance both the tourism and business profile, Ainley and Kline (2014) find farmer-to-farmer network connections are critical in agritourism capacity development.

Overall, there are several incentives and advantages of adopting agritourism, although the above barriers may limit the potential success and returns. Rural tourism promotes economic development and diversification, as well as consequent socio-cultural development, including maintenance and improvement of public services, revitalization of local crafts, customs and cultural belongings, and increased chances for social contacts and exchange, indicating positive outcomes if successful.

In summary, the literature highlights that agritourism is a possible and increasingly available option for farms looking for diversification and enhanced profitability. None the less, to access this, the farms and farmers need to have the total package of resources that tourists demand. For many farmers this requires investment in resources and skill development, which unfortunately may not be available due to the relatively isolated rural setting that makes them attractive. As such, planning and development officials have a crucial role to play in assessing agritourism potential, interest, capacity gaps, and then facilitating the filling of gaps.

2.2 Sri Lankan Agritourism

As noted earlier, in this chapter we investigate emerging economies' farmers' perceptions of agritourism, using a Sri Lankan case study. Reflecting many South Asian economies, Sri Lanka's economy was founded upon agriculture. The island country has widespread irrigation facilities, monsoon rains and fertile lands, dominated by paddy, coconut, tea and rubber (Senanayake and Wimalaratana, 2010).

None the less, with the country's liberalization, starting in 1977, and especially stability after the end of the civil war (1983–2009), further opportunities are emerging, including tourism. Tourism is the fastest-growing sector of the Sri Lankan economy, contributing 4.9% to Sri Lanka's gross domestic product, and the third-highest foreign exchange earner (US$4380.6 million in 2018) (Sri Lanka Tourism Development Authority (SLTDA), 2019). International tourist arrivals to Sri Lanka are increasing, from 0.04 million in 1970 to 0.40 million in 2000, 0.65 million in 2010, 1.03 million in 2012 and 2.33 million in 2018 (SLTDA, 2019) (Table 2.3).

Within this context, agritourism is attracting more attention. Indeed, Sri Lankan tea, spice and floral tourism have been investigated (Senanayake and Wimalaratana, 2010; Malkanthi and Routry, 2011). These studies find that while there were some agritourism ventures, and a lot of potential, not much

Table 2.3. Sri Lanka international tourist guest nights and arrivals, 2005–2018 (source: authors' own compilation based on information from SLTDA, 2019).

Year	Registered accommodation guest nights	Arrivals
2005	2,857,575	549,308
2006	3,134,243	559,603
2007	2,777,599	494,008
2008	2,763,223	438,475
2009	2,818,487	447,890
2010	4,126,544	654,476
2011	5,011,480	855,975
2012	5,038,066	1,005,605
2013	6,969,238	1,274,593
2014	7,982,110	1,527,153
2015	8,945,380	1,798,380
2016	10,595,118	2,050,832
2017	23,068,836	2,116,407
2018	25,204,996	2,333,796

action had yet taken place. Of particular interest for us are rubber farms. The interest is due to the scale (11th largest national rubber producer, 102,900 t production in 2014) (International Rubber Study Group, 2015), lack of previous agritourism investigations and its susceptibility to a range of challenges. These challenges include: downward price fluctuations (Wettasinghe, 2015); sudden release of buffer stocks by other countries (Thai PBS, 2015); changes in demand (Association for Natural Rubber Producing Countries, 2015); and reduction of tappable days (Eco-Business, 2020). Indeed, the sector is so susceptible that instead of replanting, some farmers are uprooting their plantations (Rubber Development Department (RDD), 2013). The regions most susceptible are those in the rain-fed lowlands of the south-western, southern and central parts of Sri Lanka, including the top growing districts of Kegalle (37,165 ha), Kalutara (28,765 ha) and Ratnapura (26,605 ha) (RDD, 2010). Rubber growers are exploring opportunities for alternative income sources to supplement the rubber crop. One of the common approaches has been the intercropping of fruit crops and spices. However, intercropping can be done only during the first five years of rubber cultivation (before the canopy closes). Agritourism is another opportunity, particularly with the accessibility of the main international airport at Colombo. Of note, tourism overtook rubber as the nation's fifth-placed foreign income earner in 2011 and rubber is now in seventh place (SLTDA, 2019).

Overall, there is large potential for agritourism in Sri Lanka, although still in its infancy. Agritourism could be a tool to improve rubber farmers' incomes and importantly their ability to continue farming. However, given the information on farm and farmer profiles, the resources and the incentives and barriers, how attractive is agritourism to Sri Lankan rubber farmers? Consequently, what planning constraints exist and what support is necessary to assist and facilitate agritourism development?

2.3 Method

Given Kalutara District is one of the largest producers of rubber (see Section 2.2) and is closest to the main international tourism entry point (Colombo), we selected it for the research. The Kalutara District consists of 36 Rubber Development Officer (RDO) divisions. Each division is dominated by small-scale (less than 2 ha) compared with medium-scale (more than 2 ha) rubber farms, with a ratio of 2:1. There are 24,947 rubber farmers in the Kalutara District (30% of national rubber farmers), with 28,765 ha in rubber (approximately 26% of national rubber land) (Department of Census and Statistics, 2002; RDD, 2010).

A survey was utilized to capture farmers' perceptions of agritourism as derived from the previous literature (Section 2.1) and supplemented by four focus group meetings with RDOs ($n = 2$) and rubber farmers ($n = 2$). The focus groups allowed for the identification of key local items to be included in the research. The survey included questions regarding farmers' willingness to start agritourism (seven items), suitability of their location for agritourism (eight items), their available agritourism resources (seven items), possible agritourism products (12 items), incentives (seven items) and barriers (ten items) to start agritourism, future agritourism potential (11 items), and rubber farming and demographic characteristics. Items were measured on a five-point Likert-type scale. The survey was written in Sinhala. A pre-test was held with five RDOs before finalizing the survey. The pilot study with ten rubber farmers showed the survey instrument to have adequate content and face validity.

RDOs were trained to implement the survey. Stratified random sampling was selected for the research. Rubber farmers of Kalutara District were stratified according to the 36 RDO divisions and proportionate to the 2:1 farm-scale distribution. To maintain the 2:1 ratio, four small-scale and two medium-scale farmers were selected from each of the 36 RDO divisions to have a total sample of 216.

Exploratory factor analysis was carried out to investigate the attractiveness of agritourism for farmers. The significant factors found were then inputted, along with demographic and land-related variables, into a standard ordinary least squares (OLS) regression analysis to determine cause-and-effects relationships with farmer willingness to start agritourism.

2.4 Results

2.4.1 Descriptive analysis

Most respondents were male (89%), over 36 years old (95%) and had advanced secondary school education as the highest level (95%). The majority had been farming for 15 years or more (64%) and the main source of income was relatively evenly split across rubber and non-rubber (e.g. job, business, tea planting, pension and other agriculture), as income from rubber is often not sufficient on its own. The respondents' age and education profile would indicate that there should be limited interest in agritourism (Loureiro and Jervell-Moxnes, 2004; Mnguni, 2010; Bagi and Reeder, 2012; Joo *et al.*, 2013); all the same, the diversified income streams indicate that enhanced farm profitability may be an incentive (Joo *et al.*, 2013). Certainly, average monthly income from rubber ranged from US$15 to US$6000 and, while the majority of farmers made some profit from their rubber land (64%), it is evident that they are looking for alternative sources of income (profit incentive) (Joo *et al.*, 2013; Schilling *et al.*, 2014).

Most respondents were not yet involved in agritourism (97%). With very limited involvement, there are also limited opportunities to observe the sector as an opportunity (Kunasekaran *et al.*, 2011). However, uninvolved neighbours were not a constraint in other agritourism contexts (Ollenburg and Buckley, 2007). Furthermore, those involved often only had incidental interactions with tourism practice, such as tourists and tours stopping to view roadside rubber tapping. Some had built accommodation cottages for those who come for recreational, research and writing purposes. During the focus groups with farmers and RDOs, it was recognized that non-participation might be due to lack of awareness, limited access to capital, limited infrastructure facilities available and apathy towards agritourism. While apathy might explain the limited existing tourism activity, the majority stated their willingness to start agritourism on their rubber lands (75%). Reasons for being prepared to start agritourism are presented in Table 2.4.

The main reasons for being prepared to start agritourism are farm-centric, with the known production and, as anticipated, a means to generate profits most highly rated (Joo *et al.*, 2013). The possibility of involving family members is rated the lowest (compared with Ollenburg and Buckley, 2007), indicating that the profit generation is a diversification or trade-off, as compared with supplementing rubber production. Given previous studies emphasize that agritourism provides little extra to farm incomes (Hjalager, 1996), this is an area that farmers need to be more informed about. In summary, the profitability of the respondents' rubber farms provided a strong indicative need for rural diversification. While a majority had diversified, mainly this was through alternative employment or within the agricultural sector.

Table 2.4. Willingness and reasons to start agritourism (source: authors' own computations).

Reasons	Mean	Standard deviation	% Agree or Strongly agree	*n*
Willingness to start agritourism	3.91	0.84	75	216
Kalutara being a rubber-producing district	3.88	0.75	81	216
As a profit-generating business	3.62	0.89	61	216
Better use of environment	3.57	0.80	56	216
Communication and transport facilities	3.40	0.94	58	216
As a hobby	3.07	1.15	48	215
As tourist places already exist in the district	2.97	1.09	36	215
Possibility of involving family members	2.79	1.16	35	216

2.4.2 Factor analysis

The descriptive findings provided a valuable initial insight into the respondents' perceptions. To further investigate the perceptions, and identify relationships between these, a principal axis factoring (PAF) with a varimax rotation was conducted. The PAF was completed on the 74 Likert-scale questions from the completed surveys. An examination of the Kaiser–Meyer–Olkin (KMO) measure of sampling adequacy (Kaiser, 1974) suggests that the sample is factorable (KMO = 0.821). The Bartlett's test of sphericity was 0.000. A 13-factor simple structure solution was identified (factor loadings ≥ 0.5; < 0.5 excluded), each made up of between two and five items (Table 2.5).

It is not surprising to see the factor patterns emerge, as most are indicated in the literature used to form the survey. For example, regional agritourism contributions emerge as a factor reflecting actual and desired community outcomes from previous studies (Table 2.1; Carter, 1998; Fleischer and Felsenstein, 2000; Barbieri, 2013; Gao *et al.*, 2014) and farm agritourism contributions reflects farmer's key agritourism incentives (Ollenburg and Buckley, 2007; Joo *et al.*, 2013; Gao *et al.*, 2014; Kline *et al.*, 2016). None the less, some factors provide indications for further respondent education. For example, the sixth factor, agritourism motivations, relates sustainable, profit and high potential together. While agritourism can provide for diversification and further employment opportunities (Carter, 1998; Sharpley, 2002; UCSFP, 2011; Tiraieyari and Hamzah, 2012; Barbieri *et al.*, 2016), the actual profitability and potential of agritourism are often not great, at least due to small scale and seasonality (Hjalager, 1996; Fleischer and Pizam, 1997; Joo *et al.*, 2013; Schilling *et al.*, 2014).

In other factors, the related items highlighted similarity between respondents and the literature; for example, when identifying regional and farm constraints to agritourism. For the former, the respondents related the policy context, incentives and support (or lack thereof) (Hall and Jenkins, 1997; Fleischer and Felenstein, 2000; Sharpley, 2002) with the need for shared farmer attitudes and connections (Phillip *et al.*, 2010; Ainley and Kline, 2014) and infrastructure (Roberts, 1996). For the latter, there was acknowledgement of the different skill and resource profiles needed for agritourism as compared with agriculture (Embacher, 1994; Bagi and Reeder, 2012).

All the same, the factor analysis was completed as a basis for the regression analysis. From the factors (Table 2.5), six dimensions were identified: (i) Contributions and endowment; (ii) Accessible and scenic; (iii) Motivation; (iv) Products and facilities; (v) Resources and knowledge; and (vi) Constraints.

2.4.3 Regression analysis

A multiple regression was conducted to see if the six dimensions, demographic factors (gender, age, education, farming experience and income) and other variables (such as environment, facilities, products and agricultural setting) predicted respondents' willingness to start agritourism. Using the enter method, it was found that six variables (i.e. Motivation, Environment, Education, Facilities,

Table 2.5. Farmers' agritourism perceptions factor analysis (source: authors' own computations).

Factors and components	Factor loading	Eigenvalue	Cronbach's α	No. of test measures	Macro factor	Variance explained (%)
Regional Agritourism Contributions		4.521	0.817	5	Contributions and endowment	64.11
Support from urban people	0.808					
Reduce urban migration	0.793					
Increase local food demand	0.655					
Develop rubber lands	0.582					
Improve communication	0.547					
Farm Agritourism Contributions		1.517	0.855	3		
Develop infrastructure	0.896					
Improve income	0.822					
Increase job opportunities	0.822					
Preserve Domestic Endowment		1.015	0.639	3		
Protect local tradition and culture	0.769					
Reduce early uprooting	0.748					
Increase local food sale prices	0.566					
Scenic Complements		3.835	0.807	5	Accessible and scenic	62.01
Near to waterfall	0.766					
Near to dam or lake	0.752					
Near to historical place	0.658					
Near to sea	0.645					
Near to river	0.625					
Accessible Rurality		1.126	0.752	3		
Landscape	0.859					
Easily accessible	0.821					
Short distance	0.656					
Agritourism Motivations		3.547	0.796	3	Motivation	50.67
Sustainable	0.657					
Profit-generating	0.565					
High potential	0.538					

Continued

Table 2.5. Continued.

Factors and components	Factor loading	Eigenvalue	Cronbach's α	No. of test measures	Macro factor	Variance explained (%)
Agritourism Products		4.665	0.779	4	Products and facilities	45.43
U-pick	0.767					
Agri-products	0.757					
Agricultural activities	0.752					
Rural games	0.636					
Agritourism Facilities		1.32	0.777	5		
Farmhouse	0.753					
Cooking	0.725					
Tree houses	0.671					
Tracking	0.614					
Accommodation	0.561					
Regional Agritourism Resources		6.066	0.849	5	Resources and knowledge	44.18
Rubber production	0.848					
Good climate	0.799					
Ecology	0.760					
Communication and transport	0.629					
Cooling climate	0.553					
Farm Agritourism Resources		1.849	0.723	4		
Knowledge	0.771					
Attractive place	0.745					
Private vehicle	0.645					
Extent of rubber lands	0.525					
Rubber and Local Knowledge		1.538	0.751	2		
Accessibility to information	0.527					
Rubber experience	0.514					

Regional Agritourism Constraints		4.136	0.747	5	Constraints	28.95
Reluctance to adopt tourism	0.733					
Environmental pollution	0.653					
Limited extension services	0.650					
Poor infrastructure	0.647					
Inter-regional competition	0.560					
Farm Agritourism Constraints		2.282	0.706	4		
Limited knowledge in entrepreneurship	0.781					
Limited knowledge in English	0.725					
Limited capital to finance	0.707					
Limited land availability	0.543					

Products and Agricultural setting) explained 55% of the variance in the respondents' willingness to start agritourism ($R^2 = 0.60$, adjusted $R^2 = 0.56$, F (20 195) = 14.625, $P = 0.000$). The standardized regression coefficients (β) for each predictor in the regression model are reported in Table 2.6.

As shown in Table 2.6, the motivation variable had the greatest influence on willingness to start agritourism. Overall, the regression model indicates two clear themes, which work in tandem. First are the positive influencers on willingness to start agritourism, and second the negative influencers. The first theme resonates with the agritourism literature; however, the second theme from the regression analysis is an interesting finding.

First, farmers are positively influenced on the likelihood to start if agritourism is perceived as having high potential to be a sustainable contributor to the farm and its profitability (Carter, 1998; Sharpley, 2002; Veeck *et al.*, 2006; Deller, 2010; Tiraieyari and Hamzah, 2012; Joo *et al.*, 2013; Schilling *et al.*, 2014), they live in an accessible and scenic environment (Lane, 1994; López and García, 2006; Gil Arroyo *et al.*, 2013; Gao *et al.*, 2014) and the farmer has a higher education level (Mnguni, 2010; Bagi and Reeder, 2012; Joo *et al.*, 2013). As the previous studies support, the alignment of incentives (contributor to farm profitability), broader tourism potential (accessible and scenic area) and personal capacity (higher education levels) understandably relates to higher likelihood for agritourism start-up.

Second, for the negative influencers, the lower the farmer considers the importance of the rural setting, and the less availability of products and facilities, the more likely they will start agritourism. Again, the second theme appears to contradict the key literature foundations for agritourism. On this, the results are the farmers' perceptions and willingness to start agritourism, as compared with likelihood for success. None the less, a potential interpretation of the second theme is that if a farmer perceives limited opportunities within their current resources and sector (low rural setting value, limited facilities and products), the more likely they will turn to agritourism to diversify their opportunities. The perception may be that currently limited agricultural resources limit opportunities in the current sector, as compared with the 'future' tourism sector. Reinterpreting this, if farmers perceive they have sufficient resources to stay within the agricultural sector, then there is less incentive to look for diversification

Table 2.6. Results of the regression analysis ($n = 216$) (source: authors' own computations).

Variable	β	t	Significance
Motivation	0.501	7.869	0.000*
Environment	0.267	3.533	0.001*
Education	0.126	2.556	0.011*
Facilities	–0.124	–1.938	0.054
Products	–0.166	–2.821	0.005*
Agricultural setting	–0.169	–2.042	0.042*

*$P < 0.05$.

(Nickerson *et al.*, 2001). Interpreting the second theme this way aligns with Sharpley's (2002) and Tiraieyari and Hamzah's (2012) comments that agritourism could improve the agricultural environment, economy and infrastructure, and, as Bagi and Reeder (2012) highlight, could be implemented with little effort. All the same, regions without tourism resources are not attractive to most tourists (Gannon, 1994) and this also underplays the different skill profile and tourism-specific resources are that are important for agritourism success (Embacher, 1994).

2.5 Planning Agritourism Development

With the increased pressure on rural areas, tourism, and more specifically agritourism, is an option to diversify and even revitalize regions. In this context, we investigated agritourism attractiveness for farmers, their motivations and consequently highlighted priority planning and development themes within a South Asian emerging economy context. Through gaining a better understanding of farmers' perceptions of agritourism, rural planning and development officials have a better understanding of the potential of agritourism adoption, barriers, and resource and skill needs. While the research was implemented in a region of Sri Lanka, the findings do present valuable insights into rural areas facing pressured and constrained agricultural opportunities, particularly in other South Asian emerging international tourism destinations.

Indeed, we have found much about farmers' perceptions of agritourism, and such findings provide planning and development intervention understandings. Three central understandings are evident. First, farmers facing constrained agricultural opportunities have a high level of interest in tourism. More interestingly, while potentially prompted by the research, this interest is in a context of low exposure. Second, and of particular value, it is especially farmers facing more extreme constraints who are more motivated to start agritourism, as compared with those who perceive they have the tourism resources and capacity. Furthermore, reinforcing previous studies, farmers' willingness to start agritourism is related to proximity to market (Gil Arroyo *et al.*, 2013; Gao *et al.*, 2014) and farmers with a higher education level (Bagi and Reeder, 2012; Joo *et al.*, 2013). For locations, this would likely also reflect tourist-demanded accessibility, while important to retain the rural scenic attributes that contrast with the urban attributes. Third, while development and planning officials may focus on the agritourism-resource-ready opportunities, it is these farmers who may not necessarily have the high willingness levels.

These three understandings present two challenges for tourism planning and development officials. First, those farmers who have the highest willingness to participate in agritourism are also likely to need the most resource and skill development to transition into tourism. The farmers' willingness seems to be incentivized due to limited capacity to respond to agricultural changes with their existing resources and, indicatively, the necessary agritourism resources and skills are also in limited supply. For planning and development officials the implication is that a resource and capacity-building focus will be needed for agritourism facilitation. Second, farmers appear to have greater expectations of agritourism's benefits than what it will likely be able to deliver. For planning and development officials this has the implication to educate farmers on probable agritourism investment and outcomes so that farmers can make informed decisions to participate.

This research has limitations, which also direct future research opportunities. The research was undertaken within a region of Sri Lanka and, as such, the findings could be context specific. Future research should investigate farmers' perceptions of agritourism in other South Asian contexts. Furthermore, the sample was specifically selected from rubber farmers, whereas previous studies have determined there is greater agritourism provision in sectors of tea and spices. Future research should assess if farmers' perceptions of agritourism are dependent upon their sector and/or the sector's dynamic context. Finally, this research specifically assessed farmers' perceptions of and their willingness to start agritourism. Future research should investigate if there is tourist demand and if there is an alignment between the resources available and what is demanded by tourists.

Conclusions

In constrained rural environments agritourism is an innovative diversification option. Agritourism is tourist interaction with agriculture on working farms. It has been found to have a multitude of purposes and benefits for farmers, rural communities, as well as tourists. Previous studies found that those farmers and farms that have adopted agritourism were younger and more highly educated, as well as close to markets, and participating in other development opportunities. In addition to these characteristics, a range of activities and resources are necessary to provide the agritourism experiences, mostly contrasted to those available in an urban environment. Furthermore, farms need to have people and capacity to adopt agritourism. Government and farm profitability are key incentives to adopt agritourism, while investment in resources and capacity building are the main barriers.

These characteristics and especially challenges are more imperative in South Asian emerging economies, which was the focus of the research. In Sri Lanka, agritourism is in its infancy, although it is emerging in tea and spice areas. Through implementing a survey with Kalutara rubber farmers, their willingness to start agritourism was identified. There was strong interest to start agritourism, particularly with others, utilizing the agricultural resources available. All the same, several barriers were identified, most linked to limited perceived government support. The benefits from agritourism were rated much higher and were also more numerous. These too were reflected in the factor analysis, which identified strong themes on the contributions and endowments agritourism would provide farms and regions, as well as the scenic compatibility and accessibility of the region. The regression analysis identified two key themes that explained farmers' willingness to start agritourism. These were first the potential for agritourism's positive contribution, accessibility and farmers' higher education level. The second theme was related to the farmers' existing pressured context, including less desirable setting, products and facilities. Overall, the findings present an interesting planning and development problem for South Asian rural areas: those farmers with the most interest and demand for agritourism are also likely to be those who are the least resourced to adopt the opportunity.

References

Ainley, S. and Kline, C. (2014) Moving beyond positivism: reflexive collaboration in understanding agritourism across North American boundaries. *Current Issues in Tourism* 17(5), 404–413.

Association of Natural Rubber Producing Countries (2015) *Natural Rubber Trends and Statistics, Monthly Report 7(5), May*. Association for Natural Rubber Producing Countries, Kuala Lumpur.

Bagi, F.S. and Reeder, R.J. (2012) Factors affecting farmer participation in agritourism. *Agricultural and Resource Economics Review* 41(2), 189–199.

Barbieri, C. (2013) Assessing the sustainability of agritourism in the US: a comparison between agritourism and other farm entrepreneurial ventures. *Journal of Sustainable Tourism* 21(2), 252–270.

Barbieri, C., Xu, S., Gil Arroyo, C. and Rozier Rich, S. (2016) Agritourism, farm visit, or...? A branding assessment for recreation on farms. *Journal of Travel Research* 55(8), 1094–1108.

Carter, S. (1998) Portfolio entrepreneurship in the farm sector: indigenous growth in rural areas? *Entrepreneurship and Regional Development* 10(11), 17–32.

Choo, H. and Petrick, J.F. (2014) Social interactions and intentions to revisit for agritourism service encounters. *Tourism Management* 40, 372–381.

Deller, S. (2010) Rural tourism, poverty and spatial heterogeneity. *Annals of Tourism Research* 37(1), 180–205.

Department of Census and Statistics (2002) Agricultural holdings, extent of major crops and livestock. Department of Census and Statistics Sri Lanka. Available at: http://www.statistics.gov.lk/agriculture/AllSectors/index.htm (accessed 27 June 2015).

Eco-Business (2020) Drought, deluge, disease: how should the natural rubber industry respond to climate change? Available at: https://www.eco-business.com/news/drought-deluge-disease-how-should-the-natural-rubber-industry-respond-to-climate-change/ (accessed 18 November 2020).

Embacher, H. (1994) Marketing for agri-tourism in Austria: strategy and realisation in a highly developed tourist destination. *Journal of Sustainable Tourism* 2(1–2), 61–76.

Fleischer, A. and Felsenstein, D. (2000) Support for rural tourism: does it make a difference? *Annals of Tourism Research* 27(4), 1007–1024.

Fleischer, A. and Pizam, A. (1997) Rural tourism in Israel. *Tourism Management* 18(6), 367–372.

Fratu, D. and Gruia, R. (2012) Rural tourism as an emerging sector in Bran-Moieciu-Fundata area. *Journal of EcoAgriTourism* 8(1), 32–39.

Fredrick, M. (1993) Rural tourism and economic development. *Economic Development Quarterly* 7(2), 215–224.

Gannon, A. (1994) Rural tourism as a factor in rural community economic development for economies in transition. *Journal of Sustainable Tourism* 2(1–2), 51–60.

Gao, J., Barbieri, C. and Valdivia, C. (2014) Agricultural landscape preferences: implications for agritourism development. *Journal of Travel Research* 53(3), 366–379.

Gibson, L.A. (1993) The potential for tourism in non-metropolitan areas. In: Barkely, D.L. (ed.) *Economic Adoption: Alternatives for Non-Metropolitan Areas*. Westview, Boulder, Colorado, pp. 145–164.

Gil Arroyo, C., Barbieri, C. and Rozier Rich, S. (2013) Defining agritourism: a comparative study of stakeholders' perceptions in Missouri and North Carolina. *Tourism Management* 37, 39–47.
Hall, C.M. and Jenkins, J.M. (1997) The policy dimensions of rural tourism and recreation. In: Butler, R., Hall, C.M. and Jenkins, J. (eds) *Tourism and Recreation in Rural Areas*. Wiley, New York, pp. 19–42.
Hjalager, A.-M. (1996) Agricultural diversification into tourism: evidence of a European community development programme. *Tourism Management* 17(2), 103–111.
Hoy, F. (1996) Entrepreneurship: a strategy for rural development. In: Rowley, T.D., Sears, D.W., Nelson, G.L., Reid, N.J. and Yetley, M.J. (eds) *Rural Development Research: A Foundation for Policy*. Greenwood Press, Westport, Connecticut, pp. 29–46.
International Rubber Study Group (2015) *Statistical Bulletin* 69(7–9), January–March. International Rubber Study Group, Singapore.
Joo, H.J., Khanal, A.R. and Mishra, A.K. (2013) Farmers' participation in agritourism: does it affect the bottom line? *Agricultural and Resource Economics Review* 42(3), 471–490.
Kaiser, H.F. (1974) An index of factorial simplicity. *Psychometrika* 39, 31–36.
Kline, C., Barbieri, C. and LaPan, C. (2016) The influence of agritourism on niche meats loyalty and purchasing. *Journal of Travel Research* 55(5), 643–658.
Koscak, M. (1998) Integral development of rural areas tourism, Trebnje, Slovenia. *Tourism Management* 19(1), 81–86.
Kunasekaran, P., Ramachandran, S., Yacob, M.R. and Ahmad, S. (2011) Development of farmers' perception scale on agro tourism in Cameron Highlands, Malaysia. *World Applied Sciences Journal* 12, 10–18.
Lane, B. (1994) What is rural tourism? *Journal of Sustainable Tourism* 2(1–2), 7–21.
López, E.P. and García, F.J.C. (2006) Agrotourism, sustainable tourism and ultraperipheral areas: the case of Canary Islands. *PASOS Revista de Turismo y Patrimonio Cultural* 4(1), 85–97.
Loureiro, M.L. and Jervell-Moxnes, A. (2004) Analyzing farms' participation decisions in agro-tourism activities in Norway: some welfare implications. Selected paper presented at *American Agricultural Economics Association Annual Meetings*, Denver, Colorado, August 1–4, 2000. Available at: http://ageconsearch.umn.edu/bitstream/19959/1/sp04lo02.pdf (accessed 6 November 2020).
Liu, A. (2006) Tourism in rural areas: Kedha, Malaysia. *Tourism Management* 27(1), 878–889.
Malkanthi, S.H.P. and Routry, J.K. (2011) Potential for agro tourism development: evidence from Sri Lanka. *Journal of Agricultural Sciences – Sri Lanka* 6(1), 45–58.
Mnguni, K.I. (2010) The socio-economic analysis of agritourism in two rural communities in the Limpopo Province. MSc thesis, University of South Africa, Pretoria, South Africa.
Nickerson, N.P., Black, R.J. and McCool, S.F. (2001) Agritourism: motivations behind farm/ranch business diversification. *Journal of Travel Research* 40(1), 19–26.
Ollenburg, C. and Buckley, R. (2007) Stated economic and social motivations of farm tourism operators. *Journal of Travel Research* 45(4), 444–452.
Phillip, S., Hunter, C. and Blackstock, K. (2010) A typology for defining agritourism. *Tourism Management* 31(6), 754–758.
Roberts, L. (1996) Barriers to the development of rural tourism in the Bran area of Transylvania. In: Robinson, M., Evans, N. and Callaghan, P. (eds) *Tourism and Culture: Image, Identity and Marketing*. Business Education Publishers, Sunderland, UK, pp. 185–196.
RDD (2010) *Rubber Land Census 2010*. Rubber Development Department, Colombo.
RDD (2013) *Annual Report*. Rubber Development Department, Colombo.
Schilling, B., Sullivan, K. and Komar, S. (2014) Examining the economic benefits of agritourism: the case of New Jersey. *Journal of Agriculture, Food Systems, and Community Development* 3(1), 199–214.

Senanayake, S.M.P. and Wimalaratana, W. (2010) *Agro Tourism and Rural Development in Sri Lanka with Special Reference to Nuwara Eliya District*. University of Colombo, Colombo, Sri Lanka.

Sharpley, R. (2002) Rural tourism and the challenge of tourism diversification: the case of Cyprus. *Tourism Management* 23(3), 233–244.

SLTDA (2019) Annual statistical report 2018. Sri Lanka Tourism Development Authority. Available at: https://sltda.gov.lk/storage/documents/0cb228cd03762f638bd515fe83edd453.pdf (accessed 18 November 2020).

Tew, C. and Barbieri, C. (2012) The perceived benefits of agritourism: the provider's perspective. *Tourism Management* 33(1), 215–224.

Thai PBS (2015) Rubber planters cry foul against sale of rubber from buffer stock to Singapore. *Thai PBS*, 19 May 2015. Available at: http://englishnews.thaipbs.or.th/rubber-planters-cry-foul-against-sale-of-rubber-from-buffer-stock-to-singapore (accessed 20 June 2016).

Tiraieyari, N. and Hamzah, A. (2012) Agri-tourism: potential opportunities for farmers and local communities in Malaysia. *African Journal of Agricultural Research* 6(31), 4357–4361.

UCSFP (2011) What is agritourism? University of California Small Farm Program. Available at: http://sfp.ucdavis.edu/agritourism/factsheets/what/ (accessed 18 November 2020).

Veeck, G., Che, D. and Veeck, A. (2006) America's changing farmscape: a study of agricultural tourism in Michigan. *The Professional Geographer* 58(3), 235–248.

Wettasinghe, C. (2015) Colombo rubber prices hit lowest levels since 2009. *Daily Mirror*, 1 April 2015. Available at: http://www.dailymirror.lk/68011/colombo-rubber-prices-hit-lowest-levels-since-2009 (accessed 6 November 2020).

3 History of Tourism Planning and Development in Nepal: Achievements and Challenges to Sustainable Future of Tourism

TEK B. DANGI[1]*, BIRENDRA KC[2] AND LAXMAN GAUTAM[3]

[1]*University of Southern Maine, USA;* [2]*University of North Texas, USA;* [3]*Nepal Tourism Board, Nepal*

Introduction

International tourism has developed phenomenally since World War II. Events such as the end of World War II and a better sense of peace and order globally, the introduction of the commercial jet plane for air travel, increased use of automobiles, improved wealth of nations, increased travel easements, innovations brought forward by information technology, and so on, contributed to the incredible growth of tourism (Goeldner and Ritchie, 2012). Despite a few shocks causing fluctuations, 'international tourist arrivals have shown virtually uninterrupted growth – from 25 million in 1950 to 278 million in 1980, 528 million in 1995, and 1,035 million in 2012' (United Nations World Tourism Organization (UNWTO), 2013, p. 2) and reached 1,401 million in 2018. Tourism accounts for 10% of the world's gross domestic product (GDP) and one out of ten jobs globally are created by tourism (UNWTO, 2019). Almost all the countries in the world whether developed or developing have incorporated tourism as a vehicle to boost their socio-economic development (United Nations Environment Programme and UNWTO (UNEP-UNWTO), 2005; Goodwin, 2007; Bramwell, 2011; Dangi and Jamal, 2016; Dangi and Gribb, 2018).

Historically, Nepal has laid great emphasis on tourism development due to its potential for socio-economic development through employment creation and foreign exchange earnings, as the country offers diverse natural and cultural attractions. The history of planned development in Nepal including tourism

*E-mail: tek.dangi@maine.edu

 Tourism Planning and Development in South Asia (eds D. Stylidis and B. Seetanah)
DOI: 10.1079/9781789246698.0003

started with the first five-year plan (1956–1961) after the country was opened to foreigners in 1951. Fourteen such five-year plans (some were three-year plans) have been implemented so far (the 15th is ongoing) and tourism development has always received a high priority (advocacy platform). However, the history of tourism development has faced many challenges, such as lack of sufficient infrastructure and superstructure, lack of skilled manpower, lack of sufficient and focused marketing, changes in political regimes and negative international publicity, among others (National Planning Commission (NPC), n.d.). While presenting the history of tourism development in Nepal with reference to some global initiatives, this chapter discusses the major achievements and challenges in (sustainable) tourism development and offers some recommendations, along with practical implications for policy-making bodies and practitioners.

3.1 Background

3.1.1 Origins and significance of sustainable development

The World Commission on Environment and Development (WCED) defined sustainable development as 'development that meets the needs of the present without compromising the ability of future generations to meet their own needs' (WCED, 1987, p. 8). The WCED made sustainability a prime agenda for development and sought collaboration among various agencies from governments, businesses and other private organizations.

The historic United Nations Conference on Environment and Development (UNCED) held in Rio de Janeiro in 1992, also known as the 'Earth Summit', formally wrote a constitution of sustainable development popularly known as 'Agenda 21' (Carruthers, 2005). Agenda 21 as a comprehensive programme of action (which is not legally binding but offers moral and practical suggestions) was adopted by 182 governments at the Rio UNCED. Hence, it can be said that tourism planning should form an integrated component of local, regional and national development planning. Further, the World Summit on Sustainable Development, held in 2002, developed an implementation plan for sustainable development with a focus on poverty alleviation once the link between poverty and environmental degradation became clearer (UNEP-UNWTO, 2005).

3.1.2 Emergence of sustainable tourism development

Hall *et al.* (2015) pointed out that sustainable tourism was hardly mentioned in the WCED (1987) report; however, it has become 'one of the great success stories of tourism research and knowledge transfer' (p. 2) in the aftermath. Consequently, the notion of 'sustainable tourism' seems to be ingrained in the policy and planning documents of international organizations – the World Travel & Tourism Council (WTTC), the United Nations World Tourism Organization (UNWTO) and the Earth Council (EC) – when they jointly published *Agenda 21 for the Travel and Tourism Industry: Towards Environmentally Sustainable Development* (WTTC-UNWTO-EC, 1995). The Agenda 21 'emphasized the need to make all travel and tourism businesses sustainable and detailed priority areas and objectives for governments and the tourism industry' (Dangi, 2016, p. 16). Other arguments underlining that concern for sustainability grew along with a cautionary platform (Jafari, 2001) once the negative impacts of human development on both natural and cultural environments were realized. Bramwell and Lane (1993) interpreted sustainable tourism development (STD) as a challenge to continuous growth in the 1960s with the advent of environmentalism.

Two inter-governmental organizations, The United Nations Environmental Programme (UNEP) and the UNWTO, deserve credit for publishing a report (2005) which details a historical perspective of STD and provides guidance for the future. The report mentions three pillars of sustainable development: economic sustainability, social sustainability and environmental sustainability (UNEP-UNWTO, 2005). The report defined sustainable tourism as: 'Tourism that takes full account of its current and future economic, social and environmental impacts, addressing the needs of visitors, the industry, the environment and host communities' (UNEP-UNWTO, 2005, p. 12).

Earlier definitions of STD mainly denoted alternative forms of tourism, known as community-based tourism (CBT), ecotourism, nature-based tourism, village tourism, etc., which started as alternatives to mass tourism with an emphasis on minimizing negative impacts and enhancing local economic benefits. However, the UNEP-UNWTO (2005) report emphasized the need for 'all forms of tourism in all types of destinations, including mass tourism and the various niche tourism segments' (p. 11) to follow guidelines for sustainable development of tourism.

While defining the scope of STD, Hardy *et al.* (2002) referred to Jafari's (2001) four major platforms or stages of tourism development: (i) advocacy, (ii) cautionary, (iii) adaptancy and (iv) knowledge-based tourism, which are strongly linked to the notion of STD. Advocacy referred to the earlier stage of tourism development during the 1950s–1960s, which was associated with many positive consequences and a few negative impacts. The cautionary platform referred to the late 1960s to early 1980s as a development of laissez-faire type of mass tourism which brought more negative impacts on natural and cultural attractions. The adaptancy stage during the 1980s proposed the form of tourism development based on research and recommended small-scale, community-based, alternative forms of tourism such as CBT suited to the place and which contributed to local people/economy. Finally, the knowledge-based stage of tourism development is considered to be the period between the late 1980s to the 1990s, which emphasized the objective and holistic development of tourism utilizing the best scientific knowledge that suited destination development. It can be said that the knowledge-based platform of tourism seems more relevant to guide the course of tourism development in the 21st century.

3.1.3 Origin of alternative forms of sustainable tourism development

Community-based tourism

CBT emerged alongside STD and is believed to have started 'in the early 1980s as the *sine qua non* of alternative tourism' (Weaver, 2010, p. 206). CBT is often used interchangeably with community/village tourism development. Planners of CBT focused on combatting the negative impacts of mass tourism in the global South and aiding such rural communities through grassroots development, community empowerment, resident participation and capacity building (Scheyvens, 2002). There is evidence that CBT was also practised in the global North after the 1980s, initiated either by the government organizations or by the industry (Murphy, 1985; Joppe, 1996; Dangi, 2018). A comprehensive institutional definition of CBT is: 'CBT is generally small-scale and involves interactions between visitor and host community, particularly suited to rural and regional areas. CBT is commonly understood to be managed and owned by the community, for the community' (Asker *et al.*, 2010, p. 2).

Pro-poor tourism

Pro-poor tourism (PPT) has been practised alongside STD and CBT in many developing countries. According to Goodwin (2007), PPT developed as a specialized, small-scale approach and as a means to poverty alleviation which has shown encouraging outcomes in African countries such as Gambia, South Africa, Tanzania and Egypt. A more inclusive definition of PPT is: 'Tourism that puts those people living in poverty at the top of the agenda. PPT strategies are concerned with reducing both absolute and relative poverty by providing tourism-related income opportunities for disadvantaged groups' (SNV Netherlands Development Organization and University of Hawaii, 2007, p. 10).

Ecotourism

Ecotourism developed as a form of alternative tourism (Fennell, 2001) in many parts of the world during the late 1980s and has been defined as: 'Responsible travel to natural areas that conserves the environment and improves the well-being of local people' (The International Ecotourism Society (TIES), n.d.). Ecotourism or nature-based form of tourism is becoming increasingly popular as one of the fastest-growing segments of the travel industry (KC *et al.*, 2019).

3.2 History of Tourism Planning and Development in Nepal

The history of planned development in Nepal including the tourism sector traces back to the 1950s when the first five-year plan for Nepal (1956–1961) came through a Royal Proclamation on 10 October 1955. Introduced in the context of the sheer lack of baseline data and expertise in planning, the document, however, set a foundation for the planned development in the country with subsequent formation of the Central/National Planning Commission (NPC) in 1955. Fourteen such periodic five-year plans (some three-year plans) have been implemented so far and the 15th five-year plan (2019/20–2023/24) is in its first year of implementation (NPC, n.d.). A table of the 15 periodic plans with their summarized goals/objectives in the tourism sector presents a chronological glimpse of tourism development efforts in Nepal (Table 3.1).

The review of all 15 periodic plans suggests that tourism development has received high priority in each plan. In the initial planning phase (1st to 4th

Table 3.1. Periodic Plans and tourism development goals/objectives in Nepal (source: authors' own compilation based on information from NPC).

Periodic Plan	Years	Major goals/objectives for tourism development
1st Plan (five-year)	1956–1961	'To disseminate widely in other lands information about the interests and attractions of the country, to build up travel facilities and modern hotel accommodations, ... to renovate points of major historic and cultural interest and ... to ensure full and satisfying opportunity to see, enjoy, and perhaps visit Nepal's matchless asset, the incomparable Himalayas with their flora, fauna, spas, lakes and waterfalls' (NPC, n.d., 1st Plan, p. 34)
2nd Plan (three-year)	1962–1965	To increase 110 hotel beds; to establish a national travel agency in order to facilitate tourists and increase foreign currency earnings; and to develop Pokhara valley and Lumbini as tourist destinations besides Kathmandu valley
3rd Plan (five-year)	1965–1970	Construction of Gaucher runway (current international airport), hotels capacity expansion in Kathmandu, construction of tourist hotels in Biratnagar and Pokhara; establish a customs and immigration office in Birganj; and develop Nepalese attractions films for international promotion
4th Plan (five-year)	1970–1975	To develop a tourism master plan; further expansion of Kathmandu airport and runway facilities to allow larger aircrafts/jumbo jets to land; establishment of various standard hotels in Kathmandu valley and outside; and to increase tourist stay of length from an average of 3 days to 5 days by the establishment of hotels in Pokhara, Lumbini and Namche Bazar
5th Plan (five-year)	1975–1980	To preserve the historical, cultural and geographical attractions in the country as recommended in the *Nepal Tourism Master Plan* (1972); and to generate additional foreign exchange through tourism
6th Plan (five-year)	1980–1985	To increase net foreign exchange receipts, to help improve the balance of payments; to develop tourism industry and enterprises, and to diversify employment and earning channels
7th Plan (five-year)	1985–1990[a]	To improve the balance of payments through increased foreign exchange earnings and by attracting higher-income tourists; to generate new employment opportunities; to increase tourist stay and ensure a wider distribution of tourism incomes; and endeavour to produce goods consumed by the tourism sector domestically
8th Plan (five-year)	1992–1997	To earn a larger amount of foreign exchange earnings; to expand the tourism activities to other parts of the country; to improve people's living standard and reduce poverty through tourism employment; to improve the quality of travel services and facilities; to develop efficient, safe and standardized air services; to improve coordination between tourism and other sectors of the economy; and to augment the use of local materials and services in the tourism industry

Continued

Table 3.1. Continued.

Periodic Plan	Years	Major goals/objectives for tourism development
9th Plan (five-year)	1997–2002	To establish backward and forward linkages of the tourism sector with the national economy; to establish Nepal as a premium destination in the world through effective publicity and promotion; and to enhance employment, income and foreign exchange earnings and extend the benefits from tourism down to the village level.
10th Plan (five-year)	2002–2007	To aim for sustainable development and qualitative promotion of the tourism sector; to conserve and preserve historical, cultural, religious and archaeological heritages; and to make air transportation services easily accessible, secure and reliable.
11th Plan (three-year)	2007/08–2009/10	To preserve, conserve and promote intellectual, cultural and archaeological heritages; to develop Nepal as a major tourist destination of the world; to develop healthy and competitive domestic and international air services by involving the private sector; and to formulate regulations for safe, organized, reliable and quality air services.
12th Plan (three-year)	2010/11–2012/13	To expand and diversify tourism development and to establish Nepal as a major travel destination; to employ tourism development as a means for economic development, employment generation, balanced regional development and poverty alleviation in an inclusive and equitable manner; and to consolidate national and international air connectivity.
13th Plan (three-year)	2013/14–2015/16	To develop Nepal as an attractive and safe travel destination and to enhance the contribution of tourism for national development; to increase both the number and quality of tourists to Nepal through national and international tourism promotion; and to ensure justifiable distribution of incomes from tourism.
14th Plan (three-year)	2016/17–2018/19	To increase the income levels of people with increased tourist arrivals by developing and promoting the country as an attractive and safe travel destination; and to increase and diversify tourist destinations and products.
15th Plan (five-year)	2019/20–2023/24	To develop Nepal as a safe, quality and tourist-friendly travel destination; to increase the contribution of tourism to the national economy through diversification in tourist places, products and services; and to distribute tourism benefits to people's level in an equitable manner.

[a]Restoration of multi-party democracy in 1990 led to the adoption of open, liberal, transparent and competitive market-oriented economic policies starting with the 8th Plan (NPC, n.d.).

Plans), the advocacy platform of tourism development received prominence since those plans emphasized promoting 'Nepal's matchless asset, the incomparable Himalayas with their flora, fauna, spas, lakes and waterfalls' (NPC, n.d., 1st Plan, p. 33). Other objectives included developing travel facilities and modern hotel accommodations, increasing the number of hotel beds and establishing a national travel agency to support tourists. Construction of Gaucher runway (current international airport), construction of tourist hotels in Biratnagar and Pokhara outside the Kathmandu valley, and establishment of a customs and immigration office in Birganj were also high on the agenda. It is worth mentioning that a ten-year *Nepal Tourism Master Plan* (published in 1972 during the 4th Plan) identified potential tourism hotspots beyond the Kathmandu valley such as Pokhara, Tansen, Lumbini and Chitwan and emphasized infrastructure development in those areas (Department of Tourism, 1972).

In the 5th Plan, there is a tendency towards cautionary approaches such as preservation of historical, cultural and geographical attractions in Kathmandu valley and outside the valley, which is re-emphasized in the 10th and 11th Plans. The overarching goals and objectives of tourism development throughout all plan periods were: developing and promoting Nepal as a safe and quality travel destination in the international marketplace; increasing foreign exchange earnings; creating more employment opportunities; increasing tourist duration of stay; increasing per-tourist-per-day expenditure; providing safe and reliable air services; and ensuring wider and equitable distribution of incomes from tourism (NPC, n.d.).

In 2009, the Government of Nepal (GoN) in partnership with the private-sector tourism industry published *Tourism Vision 2020* to guide tourism development throughout the country (Ministry of Culture, Tourism and Civil Aviation (MoCTCA), 2009). The Vision laid the goals of increasing annual international tourist arrivals to Nepal to two million by 2020 and of augmenting economic opportunities and increasing employment in the tourism sector to one million (MoCTCA, 2009). In 2020, the whole country was celebrating *Visit Nepal Year 2020* with a target of inviting two million international tourists; however, the GoN was forced to cancel the campaign by the end of March amidst the threat of COVID-19. *Nepal's National Tourism Strategic Plan, 2016–2025* (MoCTCA, 2016a) also identified two major long-term objectives for tourism development: (i) to develop Nepal as a major travel destination in the region, thereby increasing the contribution of tourism to the national economy; and (ii) to diversify tourism destinations inside the country taking into account the expectations and experiences of international tourists to Nepal.

3.3 Major Achievements in Tourism Development in Nepal

A strong institutional structure was created to realize the tourism planning goals through the establishment of the Department of Tourism in 1959 (dissolved to set up the Nepal Tourism Board (NTB) in 1998 with private-sector participation), the Royal Nepal Airlines Corporation (RNAC) (now NAC) in 1958, the Department of Civil Aviation in 1957 (replaced as the Civil Aviation Authority in 1998) and the Ministry of Tourism in 1978. However, the GoN reinstated the Department of Tourism in July 2014 considering its wider role. All these organizations, including the Nepal Academy of Tourism and Hotel Management (NATHM) (established in 1973 as HMTTC for tourism manpower development), collaboratively work within the policy guidelines of MoCTCA.

Tourism promotion campaigns such as *Visit Nepal Year 1998, Destination Nepal Campaign 2002–2003, Nepal Tourism Year 2011* and *Visit Nepal Year 2020* (which only lasted about two months) have augmented both national and international awareness and interest in Nepalese tourism. The first tourist arrivals data in Nepal were recorded in 1962 with 6179 tourists, which increased to 463,684 by 1998. Tourist arrivals hit the lowest number (i.e. 275,468) in 2002, mainly due to the decade-long internal Maoist conflict (1996–2006) and negative international travel advisories. The devastating earthquake of May 2015 had severe negative impacts on tourist arrivals, lowering the numbers to 538,970 in 2015, but post-earthquake recovery hit a new record of 1,173,072 in 2018 (MoCTCA, 2012, 2016b, 2018).

Reviews of various periodic plans regarding tourism development highlight the progress made during those plan periods. A review of the 1st Plan identified inadequate air transport and hotel facilities as the major bottlenecks to tourism development, which were greatly overcome due to the efforts made during the 2nd Plan period (as stated in the 2nd Plan review). The review of the 3rd Plan stated setting up about 800 hotel beds of varying standards in Kathmandu, the establishment of different travel agencies, continued expansion and development of Tribhuvan airport and initiation of foreign airlines' services to Nepal; and the national air carrier's (then RNAC's) increase in the international services played a major role to tourism development in Nepal in the preceding decade. Other factors contributing to tourism development included the development of places like Nagarkot and Kakani as tourist attractions, availability of sightseeing services, trained guides and growing international publicity for Nepal undertaken by the Department of Tourism (NPC, n.d.). Some of the major achievements made in the 5th Plan included that the number of hotel beds increased from 1663 in 1974/75 to an estimate of 6838 in 1979/80. In the 5th Plan period, 1095 persons were trained with various skills as required for the tourism sector. The Nepal Industrial Development Corporation invested 115.4 million Nepalese rupees in tourism-related industries during the 5th Plan period. Further, in the 6th Plan, permission was granted to international charter flights under the private sector to enhance tourism arrivals. The 7th Plan review acknowledged that the contributions of the tourism sector in overall national development remained quite significant, including the balance of payments through foreign exchange earnings.

The review of the 8th Plan highlighted that extension of trekking, mountaineering, national parks and other places of tourist attractions outside the Kathmandu valley contributed significantly to increasing the number of tourists visiting the country and to diversifying tourism beyond one central location – the Kathmandu Valley. The review also noted the significant foreign exchange earnings from tourism, stating 'while the earning from the tourist sector was only US$78,000 in 1961/62, US$68,343,000 was earned in 1989/90 with an average increase of 26% per annum' (NPC, n.d., 9th Plan, p. 9). Opening up of new trekking routes in Mustang and Manang regions for guided trekking (in 1990/91 and 1991/92), granting permission to three private companies to operate air services and to one to operate a helicopter service took place in the 8th Plan. Further, the installation of terminal tower equipment in Tribhuvan

International Airport, Nepalgunj and other airfields was the major achievement of the 8th Plan. The Tourism Infrastructure Development Project, launched during the 8th Plan with funding from the Asian Development Bank, completed some major works including Pokhara conservation area improvement, Ghale Gaun–Sikles ecotourism development, Pokhara airport runway improvement and completion of the tourist service centre buildings in Kathmandu and Pokhara. During the 9th Plan, successful launching of *Visit Nepal Year 1998* played an important role in creating tourism awareness in the country and promoting Nepal internationally, drawing attention and local investment in tourism, developing and diversifying tourism destinations inside the country through organizing dozens of festivals and events, and reaching aviation agreements with closely tied countries (NPC, n.d.).

The peaceful settlement reached between the Maoists and GoN in November 2006 ended the decade-long internal conflict in Nepal giving new hopes for socio-economic development including tourism. Some of the major achievements made during the 10th Plan include removal of heritage sites in Kathmandu valley from the United Nations Educational, Scientific and Cultural Organization's (UNESCO's) endangered sites list and successful conclusion of *Destination Nepal 2002–03*, which focused on regional tourism promotion in the aftermath of the 9/11 incident in the USA that was followed by a sharp decline in international arrivals. An international-standard Mountain Academy of Nepal was established, and the Golden Jubilee Celebration of Mount Everest was observed in 2003 amidst the presence of world-renowned mountaineers and international media. Nepal's first professionally designed marketing slogan – 'Naturally Nepal: Once is not enough' – was successfully launched in 2006. During the 10th plan, NATHM started offering bachelor's courses in Hotel Management and Travel and Tourism Management followed by a master's in Hospitality Management during the 12th Plan period.

With a view of restoring Nepal's pristine image in the international marketplace, which was significantly tarnished due to the decade-long internal Maoist conflict, the country observed *Nepal Tourism Year 2011* with a goal of attracting one million tourists. Unfortunately, *Nepal Tourism Year 2011* failed to meet the target as only 736,215 tourists visited Nepal in 2011. However, tourism has increased during the post-conflict era, giving hope for the future. Causes of such failures have been discussed under challenges in aggregate. Moreover, the 15th Plan starts with an optimistic note that formation of all three tier governments based on federal structure, fast-moving reconstruction of earthquake-damaged cultural structures and development of new tourism destinations are taking place, and that construction of the second international airport is gaining momentum, creating a more conducive environment for tourism development (NPC, n.d., 15th Plan).

3.4 Nepalese Experience in Sustainable Tourism Development

Nepal as a travel destination has practised various forms of STD such as CBT, PPT, mountain tourism and ecotourism/nature-based tourism (detailed below) as a means to employment creation, foreign exchange earnings, preservation of natural and cultural heritage, poverty alleviation and livelihood improvement of local people, supported by government and various international agencies. Achievements and challenges faced by Nepal in STD need a debate and discussion to guide a better future course of action.

3.4.1 Mountain tourism

Any description of Nepalese tourism will be incomplete without a mention of mountain tourism. After the first successful ascent of Mount Everest on 29 May 1953 by Tenzing Norgay Sherpa of Nepal and Sir Edmund Hillary from New Zealand, Nepal became the centre of mountaineering and trekking activities. Nepal boasts eight of the 14 mountains higher than 8000 m in the world, including Mount Everest. As of 2014, GoN has opened 414 peaks for mountaineering activities. In 2018 alone, a total of 3136 climbing permits were issued and 1603 climbers made successful ascents of different mountain peaks in Nepal. Further, in 2018, a total of 187,692 foreigners (16% of total tourists) visited Nepal for trekking and mountaineering purposes (MoCTCA, 2018).

The GoN has shown due concern for the protection and preservation of the mountain environment and culture by enforcing befitting regulations. The mountaineering regulation requires each expedition team to obtain a climbing permit and the regulation states: 'The team shall not damage or destroy any tree, shrub, other forest, heritage or environment existing in the adjoining area or on the route of its mountaineering expedition' (Department of Tourism, 2018, p. 342). Further, the regulation states: 'The team shall not indulge in any activity that may be contrary to the customs and religious, social or cultural traditions of the local people of that area' (p. 342). For the management of garbage, the regulation classifies garbage used by the mountaineering expedition teams into three categories (destroyable, recyclable and re-exportable) and makes mandatory provisions for expedition teams to manage garbage as specified by the regulation.

Further, frequent clean-up expedition teams launched by the GoN in partnership with local agencies such as the Nepal Mountaineering Association and foreign expedition teams on Mount Everest and other mountains, as well as the measures mentioned above, exemplify some of the serious endeavours taken by the GoN for sustainable mountain tourism development. The Mountaineering Royalty Disbursement and Operation Guideline (MoCTCA, 2013) provisions that the government will share 30% of its mountain royalties with local bodies (District Development Committee/s) for the development and maintenance of tourism-related products and facilities in the given area. The spirit of central–local government partnership for tourism development through the equitable distribution of tourism benefits seems to follow some of the guiding principles of justice tourism (SNV Netherlands Development Organization and University of Hawaii, 2007; Jamal *et al.*, 2013; Dangi, 2016) and present an example of the knowledge-based platform in practice.

3.4.2 Village tourism/homestay as a form of community-based tourism

The history of CBT in Nepal started with the launching of Sirubari Village Tourism/Homestay Tourism in Syangja district in western Nepal in 1997. It was initiated and marketed by Nepal Village Resorts (a Kathmandu-based travel agency) in partnership with the local community, which is predominantly ethnic Gurungs and other occupational castes consisting of Damai, Kami and Sarki. Sirubari homestay provides visitors with a local-style accommodation (mostly owned and managed by ethnic Gurungs), locally produced home-made food and engaged community activities. The Tourism Development and Management Committee, mainly composed of the local community representatives, assigns visitors to different hosts on a rotational basis to ensure equitable benefits from tourists.

The major tourism resources of the village are local Gurung culture demonstrating a unique ethnicity and livelihood patterns, a community forest and the Thumro Juro viewpoint. The first commercial group of 16 Belgian tourists visited Sirubari in October 1997. It reached the highest number of 1521 international and 287 domestic tourists (total 1808) in 2000/01, which declined to 524 international and 33 domestic tourists (total 557) in 2009/10 mainly due to the internal Maoist conflict (Thapa, 2010).

As a result of the peace accord signed between the government and the then-revolutionary Maoists in 2006 and followed by the promulgation of the new constitution of Nepal in 2015 and the formation of a stable government, tourism development has gained momentum again. Recognizing the popularity and potential of village tourism to contribute to the local economy and to improve the quality of living through homestay tourism-related jobs, the GoN formally introduced the *Homestay Sanchalan Karyabidhi, 2067* (MoCTCA, 2010). This directive provides detailed guidelines on how to establish, manage and promote homestay tourism in Nepal. Many rural places such as Bandipur in Tanahu, Ghalegaon in Lamjung, Ghandruk in Kaski, Bridim in Rasuwa and Barpak in Gorkha have significantly increased the number of registered homestays (community and private houses) in various districts of Nepal, which reached 324 in 2018 (MoCTCA, 2018).

3.4.3 Pro-poor tourism: a case study

The Tourism for Rural Poverty Alleviation Programme (TRPAP), implemented as a five-year CBT programme (September 2001–2006), was funded and supported by the United Nations Development Programme (UNDP), the Netherlands Development Organization (SNV) and the UK Department of International Development (DFID) in partnership with GoN, NTB, other government and non-government organizations and the private sector. The programme carried out its pilot projects in six touristic districts of Nepal, namely Taplejung, Solukhumbu, Rasuwa, Chitwan, Rupandehi and Dolpa, a total of 48 Village Development Committees, covering 715 settlements, over 28,000 households and around 160,000 expected beneficiaries. The programme aimed to contribute to poverty alleviation in Nepal by mainstreaming pro-poor sustainable tourism policies and developing strategies and innovative models that were pro-women, pro-environment and pro-communities (TRPAP, 2007).

TRPAP achieved replication-worthy results in alternative livelihoods capacity building, skills development training, increased agricultural productivity, increased local employment opportunities, empowerment and the improved ability to combat poverty. TRPAP was also acknowledged for small-scale rural tourism infrastructure development initiatives such as the construction of trails, bridges, information centres, and improved water and sanitation. As part of tourism infrastructure development, seven new tourism products were developed. It formed and re-formed 781 Community Organizations with 15,977 members with women having a share of 50.65% (TRPAP, 2007).

Through venture capital funds, the project provided soft loans at favourable interest rates only to the members of TRPAP for small entrepreneurs, thus promoting local skills like handicrafts, agri-products, fisheries, poultry farming, livestock rearing, hotels, restaurants and homestays, trekking guides training and much more. Grant preference was given to the poor, women and disadvantaged groups, which were channelled through the Sustainable Tourism Village Fund (STVF). However, the TRPAP district sites faced problems during implementation such as a lack of elected representatives as elections could not be held due to the internal Maoist conflict (TRPAP, 2007), including low tourist arrivals in the project districts. Skill drain and brain drain were big threats to tourism entrepreneurship.

3.4.4 Nature-based tourism

Nepal has significantly improved its natural resource conservation strategy over the years, with 40% of the land being covered by forests (KC *et al.*, 2014). Chitwan National Park (NP), established in 1973, is the first national park in Nepal. Since then, the protected area (PA) coverage has grown considerably. Currently, there are 12 national parks, one wildlife reserve, one hunting reserve and six conservation areas. About 23% of the land coverage is allocated to PAs (Department of National Park and Wildlife Conservation, 2018).

With the increased emphasis on conservation, there is also a growing interest and popularity in PA-based nature-based tourism opportunities. Even though not all the PAs are accessible to tourists (hence remain mostly unexplored) due to their geographical locations, a handful of PAs get significant attention and visitation from international tourists. Among the PAs, Annapurna Conservation Area, Chitwan NP, Sagarmatha (Mount Everest) NP and Shivapuri Nagarjun NP are the most visited PAs in Nepal. In particular, Chitwan NP and Sagarmatha NP were listed under UNESCO World Heritage Sites in 1979 and 1984, respectively (UNESCO World Heritage Center, n.d.), and attract visitors due to added value.

PAs are promoted for both 'protection' as well as 'use' functions (Butzmann and Job, 2017). Tourism development in the form of 'ecotourism' is very common in PAs in Nepal. A few PAs are doing more than good in tapping this nature-based tourism niche market both for preservation and economic development. PA-based tourism is very much consistent with the concept of STD, whereas basic principles of CBT, PPT and TRPAP are embedded within the PA-based tourism initiatives. However, there are many challenges along with the opportunities for PA-based tourism. Although the popularity of eco-tourism/nature-based tourism is a positive asset for Nepal, it is time to be more aspiring and to promote sustainable thinking and practices. There are major ongoing concerns and debates about how to prevent over-tourism in some of these PAs (e.g. Mount Everest NP and Chitwan NP). Fair distribution of tourists to these pristine areas is rather important and needs more strategic thought process from experts. Ecotourism projects in communities near PAs do create positive economic and environmental impacts (Nyaupane and Poudel, 2011) and the outcomes from ecotourism projects tend to be contextual (KC *et al.*, 2018). However, with adequate planning and proper governance structure, eco-tourism/nature-based tourism potential in Nepal can be harnessed effectively.

3.5 Challenges in Sustainable Tourism Development

Despite major collective efforts mentioned earlier and given the unique nature of Nepal's tourism attractions, Nepalese tourism seems short of achieving its full potential due to several internal and external challenges. The review of the 1st Plan acknowledged inadequate air transport and hotel facilities as the major bottlenecks in the very early stage. Further, the review of the 6th Plan (after 30 years of planned development) identified that for many years, the benefits of the tourism sector were confined to a few pockets of the country. The review also acknowledged that 'there have been no significant successes in expanding the role of this sector in an organized and well-planned manner to realize the potential benefits in the overall national development endeavors. Even the increase in the flow of tourists to other areas outside Kathmandu has not yielded the expected benefits to the general mass' (NPC, n.d., 7th Plan, p. 349). The review also puts a warning sign (cautionary platform) that the inflow of tourists in limited attractions (lacking infrastructure development in other areas) for a relatively long period contributed to the consequent decline of those attractions.

The 8th Plan review reiterated that lack of infrastructure hampered the proper development and expansion of places of tourist attractions in the country, which negatively affected tourists' duration of stay and per tourist per day expenditure. The 8th Plan envisioned that construction of a second international airport will be initiated once a detailed study is completed with the selection of the most suitable site; however, the construction of the second international airport in Nijgadh, Simara has been plagued by controversies relating to the clearance of about 2.4 million hardwood forest (Sal) trees as of 2019.

The 9th Plan described other challenges such as confinement of tourism's share of the country's GDP to 4.4%, non-integration of tourism with other socio-economic sectors in the country, economic returns from tourism confined to a few geographical regions and communities, shortage of basic infrastructure, ecological degradation, fragile institutional arrangements, incompetent human resources, lack of managerial capability, and inadequacy of quality tourism products and services. Limited international air-seat capacity, lack of competitive international tourism marketing, inadequate tourism activities to increase tourist duration of stay, inadequate retention of tourism receipts and low per tourist per day expenditure have been mentioned as additional challenges (NPC, n.d., 9th Plan). The review of the 9th Plan acknowledged that the success and enthusiasm created by the *Visit Nepal Year 1998* campaign were slowed due to internal peace and security issues created by the internal Maoist conflict.

The 10th Plan delineated internal and external challenges, which included: suspension of Indian Airlines' flights following the hijacking of one of its flights (on 24 December 1999) that took off from Kathmandu for India; the tragic royal palace massacre of June 2001; the 9/11 terrorist attack in the USA followed by the Afghanistan War; constraints in air seats from European sectors following RNAC's termination of European flights; and reduction of flights by other international airlines. Lack of a tourism master plan for integrated tourism development and continuing the achievements gained by TRPAP remained as additional challenges.

A review of the 11th Plan found the lengthening of the political transitional period (recurring in 13th Plan review), lack of tourism investment and frequent shutdowns (mostly by political parties) mainly responsible for obstructing tourism development. A review of the 12th Plan reiterated issues such as lack of reliable air services, lack of justifiable distribution of tourism benefits (recurring in 14th Plan review), lack of integrated development of tourism infrastructure and facilities (recurring in 13th and 14th Plan reviews), failure to correct the negative international publicity of Nepal and failure to find a strategic partner for the national carrier (NAC) to improve its management and services. A review of the 13th Plan added that severe impacts of the devastating 2015 earthquake on infrastructure including World Heritage Sites negatively affected the inflow of tourists to Nepal in the succeeding years. Further, major challenges identified in the review of the 14th Plan included lack of proper diversification of tourism products, lack of proper international marketing and promotion, lack of proper involvement of the private sector in the preservation and promotion of religious and cultural sites, and failure of tourism to embrace modern technology in its development (NPC, n.d., 14th Plan).

The ongoing 15th Plan summarized the major challenges facing the tourism industry, which included to establish Nepal as a premium travel destination in the world, to identify new tourist destinations inside the country for tourism diversification, to entice private-sector investment for tourism infrastructure development, to develop sufficient skilled manpower required for tourism and to minimize the impacts of global climate change. Other challenges to be addressed included increasing the length of stay and expenditure of tourists, attracting high-spending (quality) tourists, and ensuring efficient and tourist-friendly services. In the context of achievements made and challenges faced by Nepal towards STD, the following future directions are proposed.

3.6 Future Directions towards Sustainable Tourism Development

Several challenges mentioned earlier in the review of different periodic plans provide a direction on what is missing and what needs to be undertaken. Further, strengths, weaknesses, opportunities and threats (SWOT) analysis of Nepalese tourism conducted while developing *Tourism Vision 2020* still holds true to provide continued direction to Nepalese tourism development. *Tourism Vision 2020* outlined opportunities such as growing outbound tourism from immediate neighbours, spillover effects from adjoining destinations of India and China, and growing demand for adventure travel (MoCTCA, 2009, p. 7). The reality is that those opportunities will always exist for Nepal. Therefore, it is pivotal for Nepal to analyse the costs, constraints and benefits of tourism development to move forward.

3.6.1 Need for infrastructure development for tourism diversification

A review of the 8th Plan and others blamed lack of infrastructure development as a barrier to tourism diversification and underlined the need for developing the tourism sector not only as a medium of foreign exchange earnings but also as a catalyst for the whole development process. Inadequate infrastructure has contributed to over-tourism in certain destinations and a lack of tourists to other destinations within the country. Without proactive government initiatives for infrastructure development, private sectors and other entrepreneurial ventures can be reluctant to join the tourism industry. Previous research findings assert that the tourist inflows into a destination and their number are influenced by infrastructure development, particularly transport infrastructure (Khadaroo and Seetanah, 2007). As a developing country, Nepal still struggles to provide proper land transportation including air transport as it does not have a strong national carrier to provide better access to tourists. Finding a strategic partner for the national carrier, NAC (as mentioned in the review of the 12th Plan), is long overdue. Examples from regional carriers such as Thai Airways, Singapore Airlines and Qatar Airways significantly contributing to destination marketing and promotion could be a great learning lesson for Nepal. With neighbouring Asian countries such as China, Thailand and Malaysia being so successful in retaining a significant portion of the international tourists, there is a huge potential for Nepal to capture this market segment; however, with a proper strategy.

3.6.2 Need for understanding the tourism markets

The UNWTO (2013) predicted that international tourism will grow by 3.3% annually from 2010 to reach 1.8 billion by 2030. Such positive trends carry significant implications for growing destinations. For Nepal, it is important to understand the market characteristics and destination preferences to create tourism demands. First, the majority of international tourists assume that Nepal is an affordable and inexpensive tourism destination. Is it affordable and inexpensive and is that conveyed to consumers? Second, tourists to Nepal tend to be more adventurous with fewer sophistication seekers; are these specific tourist segments being targeted yet? Do concerned authorities and major stakeholders understand this perceived image of Nepal among international tourists? Have Nepal's tourism marketing efforts been tied to outbound market trends in generating markets and do marketing plans achieve targets? Based on the tourist arrivals to Nepal, its growth rate and the target goal, it is clear that Nepal is still trying to figure out its consumers. Thus, Nepal perhaps lacks the focused marketing effort (as pointed out in the reviews of the 9th and 14th Plans) to reach out to its potential market segment. Mostly, international tourists to Nepal are based on word-of-mouth advertisement. Promotion through digital marketing and social media has not been that visible. Based on the statistics, Nepal seems more focused on bringing international tourists, and domestic tourism seems somehow neglected as there are no statistics or data management systems to monitor domestic tourism. It is important to create a mechanism and capture this growing domestic market segment in a planned manner besides promoting Nepal for international tourists.

3.6.3 Need for promotion beyond limited predictable destinations

In Nepal, most of the tourist destinations are becoming predictable. Some of the most popular destinations for tourists include the capital (Kathmandu valley), a few national parks (Chitwan, Mount Everest and Shivapuri Nagarjun), a few other cities (Pokhara, Nagarkot, Lumbini) and a few mountain-based destinations (Annapurna Conservation Area, Mount Everest Base Camp, etc.). Due to the popularity of these limited destinations, often these destinations are experiencing over-tourism. For instance, over-tourism on Mount Everest circulated all over social media in 2019. Likewise, many destinations in the list mentioned above are going through such over-tourism concerns, at least during the peak tourist season. Even now, Nepal is struggling to promote tourism throughout the country on a year-round basis to minimize over-tourism in certain destinations. At this point, Nepal also needs to diversify the attractions and opportunities, and perhaps promote untapped destinations, to mitigate overcrowding in limited tourism destinations. Such a tourism development policy will also address the unbalanced regional economic development. However, as mentioned earlier, the lack of proper infrastructure development is a major obstacle to fostering tourism development in untapped destinations.

3.6.4 Need for an effective and integrated approach to tourism development

Nepal possesses significant potential for economic development as the country offers diverse opportunities for tourists with different preferences such as cultural or nature-based tourists. With this potential, Nepal has prospects to address the United Nations' Sustainable Development Goals (SDGs) (e.g. support poverty reduction; ensure inclusiveness and equality; promote sustained, inclusive and sustainable economic growth; protect, restore and promote sustainable use of terrestrial ecosystems) via tourism development (UNWTO, 2018). However, all stakeholders need to collaborate to optimize the benefits of tourism. Once over-tourism becomes an issue for a destination, its negative impacts can burden the destination, which makes it difficult to recover from the over-tourism effect. Most importantly, the government and the tourism industry can and should play a pivotal role to shape tourism. For instance, the GoN can take authoritative decisions to devise and implement improved policy and planning to promote sustainable practices in tourism destinations. The tourism industry itself should be more sustainability focused to thrive amidst burgeoning tourism trends while being competitive through the adoption of sustainable practices. Both the government and the tourism industry should focus on sharing the tourism benefits with the host communities and residents to foster community well-being and garner their support for tourism development. The tourism industry is a dynamic industry that is rapidly changing in terms of growth, destination development, tourism market preferences, diversification of tourism products, etc. Even though Nepal needs a new tourism master plan to reflect the changes in the tourism industry, it also needs a tactical plan (i.e. short-term plan) to be adaptive. *Nepal's National Tourism Strategic Plan, 2016–2025* fulfils this need to some extent; however, its true implementation outcomes are still awaited. Tackling the impacts of global climate change on the Nepalese Himalayas including tourism also needs immediate attention.

Further, having sufficient skilled manpower in tourism can be an equally crucial aspect to support the intended outcome. Most importantly, the socio-political system has gone through many changes in regime while creating absolute instability within and for other sectors overall. Relationships with stakeholders and their active participation within the tourism sector are vital, but other cognate sectors including tourism education and training operations concerning the tourism sector can determine its success as well.

3.6.5 Need for an aggressive and outcome-oriented destination marketing organization

Realizing the need for a public–private partnership organization for professional and competitive destination marketing, the NTB was established in December 1998 by dissolving the then government-run Department of Tourism. Despite a limited budget and staff and negative international media advisory in the wake of the Maoist conflict in the country, the Board within its first five to ten years was able to develop as a strong and model institution of public–private partnership in the country and abroad and to display its worth through effective destination marketing and promotion. The first author had the opportunity of serving on the Board first as marketing director and then as CEO for four years. Realizing the uncertainty of funding collected by NTB through 2% Tourist Service Fees (from different tourism establishments such as hotels and restaurants) and the relatively small budget, the first author (as the CEO) successfully lobbied the GoN (through making a recommendation to Dr Khatiwada's Task Force on Tourism) to change the Tourist Service Fees collection mechanism, adding it into the tourist airport departure tax. As a consequence of the Tourist Service Fees collection mechanism change, the NTB's budget increased to almost US$2.5 million in 2007 from around US$1 million in 2004. Currently, due to the increase in the Tourist Service Fees collection rate coupled with an increase in tourist arrivals, the NTB's annual budget stands at around US$10 million and it has added staff to function (pre-COVID-19 scenario). However, given the political stability achieved by the country coupled with a conducive environment, the current overall functioning of the NTB including international marketing and promotion does not seem to be effective and fulfilling its mission. A review of the 14th Plan pointed to the lack of proper international marketing and promotion as a major challenge to tourism. Without a strong, professional destination marketing organization (e.g. like the Tourism Authority of Thailand (TAT) and Singapore Tourism Board (STB)), no travel destination has achieved its target of the desired number of tourists and tourism earnings. The NTB was established to emulate the exemplary destination marketing organizations such as TAT and STB in the region. Now, the GoN possibly needs to consider re-evaluating, revamping and restructuring the national tourism organization to tie its activities to outcomes.

Conclusions

Nepal has emphasized the tourism industry's potential for the socio-economic development of the nation since the industry's inception in the 1950s. Several five- and three-year plans were laid out to achieve STD with significant success but largely failed to reach the expected outcomes owing to several internal and external challenges. From the 1st Five-Year Plan (1956–1961) until the 15th Five-Year Plan (2019/20–2023/24), there has been a consistent focus on increasing tourism revenue, diversifying tourism products, enhancing infrastructure, and establishing Nepal's brand image as an attractive and safe travel destination. Of late, revenue from tourism has been viewed closely with the overall societal well-being via equitable distribution of benefits from tourism. In the history of (sustainable) tourism development, Nepal seems to follow the popular tourism trends and experiments happening on a global scale. The four platforms of tourism development (advocacy, cautionary, adaptancy and knowledge-based) have each affected the course of Nepal's tourism policy and programme implementation to some extent. Adaptancy, or the alternative form of tourism popularized in the 1980s, has been widely practised in Nepal mainly in the forms of CBT, along with PPT, PA-based tourism/ecotourism and mountain tourism. We conclude that Nepal still struggles to provide adequate infrastructure for tourism development, causing overutilization or underutilization of tourism resources in the country. Additional challenges in achieving expected goals, at times, are triggered by internal conflicts followed by lengthy political transitioning. Nepal offers unique tourism products given its topography and biodiversity, which require proper planning and strategic marketing approaches. Considering the continued lack of proper infrastructure, lack of understanding of its tourism market and perhaps the Nepalese tourism sector trying to achieve its goal without required cooperation and coordination with other cognate sectors, it calls for an effective and integrated approach to tourism development following the successful models in the region and applying the knowledge-based platform of tourism.

Only secondary data were utilized in this chapter to glean a comprehensive review of tourism planning efforts, achievements and challenges, as well as the future potential of tourism development, in Nepal. Empirical studies can further enhance the contextual understanding of the Nepalese tourism industry. Hence, future studies should gain broader perspectives by incorporating key current and past tourism stakeholders and drivers of tourism in the government (such as MoCTCA, NPC) as well as leaders of private travel–trade organizations including other cognate sectors of the industry. Given the infrastructure development issues causing difficulty exploring many potential destinations and overcrowding at a few popular destinations within Nepal, focused research needs to be undertaken in the future to examine how such issues could be properly addressed from the perspectives of sustainable tourism planning. For many years, Nepal has been trying to increase the number of international tourist arrivals. Even though Nepal holds a special place for the international tourism market, it is equally important to define, emphasize and facilitate the domestic tourism market. Indeed, prospects for the domestic tourism market and its growth are something that needs to be explored in shaping the Nepalese travel and tourism industry, as its importance seems more prominent in the post-COVID-19 tourism scenario. This study attempted to link STD efforts in Nepal with SDGs in general; however, more in-depth assessment is required to evaluate how such efforts could be directed towards achieving SDGs through tourism.

References

Asker, S.A., Boronyak, L.J., Carrard, N.R. and Paddon, M. (2010) *Effective Community Based Tourism: A Best Practice Manual*. APEC Tourism Working Group, Asia-Pacific Economic Cooperation, Singapore.

Bramwell, B. (2011) Governance, the state and sustainable tourism: a political economy approach. *Journal of Sustainable Tourism* 19(4–5), 459–477.

Bramwell, B. and Lane, B. (1993) Sustainable tourism: an evolving global approach. *Journal of Sustainable Tourism* 1(1), 1–5.

Butzmann, E. and Job, H. (2017) Developing a typology of sustainable protected area tourism products. *Journal of Sustainable Tourism* 25(12), 1736–1755.

Carruthers, D. (2005) From opposition to orthodoxy: the remaking of sustainable development. In: Dryzek, J.S. and Scholsberg, D. (eds) *Debating the Earth: The Environmental Politics Reader*. Oxford University Press, New York, pp. 285–300.

Dangi, T.B. (2016) Towards a robust framework of sustainable community-based tourism (SCBT): exploring destination justice and equity as a part of governance, a case study of Bryan-College Station, Texas, USA. Doctoral dissertation, Texas A&M University, College Station, Texas.

Dangi, T. (2018) Exploring the intersections of emotional solidarity and ethic of care: an analysis of their synergistic contributions to sustainable community tourism development. *Sustainability* 10(8), 2713.

Dangi, T.B. and Gribb, W.J. (2018) Sustainable ecotourism management and visitor experiences: managing conflicting perspectives in Rocky Mountain National Park, USA. *Journal of Ecotourism* 17(3), 338–358.

Dangi, T.B. and Jamal, T. (2016) An integrated approach to 'sustainable community-based tourism'. *Sustainability* 8(5), 475.

Department of National Park and Wildlife Conservation (2018) Annual Report. Available at: https://drive.google.com/file/d/1HUBzp9WiitBEX0eyx829RSN8ng4oa_IR/view (accessed 15 October 2019).

Department of Tourism (1972) *Nepal Tourism Master Plan*. Government of Nepal, Kathmandu.

Department of Tourism (2018) Mountaineering in Nepal: Facts and Figures. Available at: https://drive.google.com/file/d/1oCZ823pRV7HgKnWhrzwPUdf9zpr71sEs/view (accessed 22 December 2019).

Fennell, D.A. (2001) A content analysis of ecotourism definitions. *Current Issues in Tourism* 4(5), 403–421.

Goeldner, C.R. and Ritchie, J.R.B. (2012) *Tourism: Principles, Practices, Philosophies*. Wiley, Hoboken, New Jersey.

Goodwin, H. (2007) Indigenous tourism and poverty reduction. In: Butler, R. and Hinch, T. (eds) *Tourism and Indigenous Peoples: Issues and Implications*. Elsevier, Oxford, pp. 84–94.

Hall, C.M., Gossling, S. and Scott, D. (eds) (2015) Tourism and sustainability: an introduction. In: *The Routledge Handbook of Tourism and Sustainability*. Routledge, New York, pp. 1–12.

Hardy, A., Beeton, R.J. and Pearson, L. (2002) Sustainable tourism: an overview of the concept and its position in relation to conceptualizations of tourism. *Journal of Sustainable Tourism* 10(6), 475–496.

Jafari, J. (2001) The scientification of tourism. In: Smith, V. and Brent, M. (eds) *Hosts and Guests Revisited: Tourism Issues of the 21st Century*. Cognizant Communication Corporation, New York, pp. 28–41.

Jamal, T., Camargo, B.A. and Wilson, E. (2013) Critical omissions and new directions for sustainable tourism: a situated macro–micro approach. *Sustainability* 5(11), 4594–4613.

Joppe, M. (1996) Sustainable community tourism development revisited. *Tourism Management* 17(7), 475–479.

KC, B., Stainback, G.A. and Chhetri, B.B.K. (2014) Community users' and experts' perspective on community forestry in Nepal: a SWOT-AHP analysis. *Forests, Trees and Livelihoods* 23(4), 217–231.

KC, B., Paudyal, R. and Neupane, S.S. (2018) Residents' perspectives of a newly developed ecotourism project: an assessment of effectiveness through the lens of an importance–performance analysis. *Asia Pacific Journal of Tourism Research* 23(6), 560–572.

KC, B., Morais, D.B., Peterson, M.N., Seekamp, E. and Smith, J.W. (2019) Social network analysis of wildlife tourism microentrepreneurial network. *Tourism and Hospitality Research* 19(2), 158–169.

Khadaroo, J. and Seetanah, B. (2007) Transport infrastructure and tourism development. *Annals of Tourism Research* 34(4), 1021–1032.

MoCTCA (2009) *Tourism Vision 2020*. Ministry of Culture, Tourism and Civil Aviation, Government of Nepal. Available at: http://old.tourism.gov.np/downloadfile/1Tourism-Vision-2020-Final2.pdf (accessed 28 December 2019).

MoCTCA (2010) Homestay Operation Directive. Ministry of Culture, Tourism and Civil Aviation, Government of Nepal. Available at: https://www.tourism.gov.np/publications/5 (accessed 16 November 2020).

MoCTCA (2012) *Nepal Tourism Statistics 2011*. Ministry of Culture, Tourism and Civil Aviation, Government of Nepal, Kathmandu.

MoCTCA (2013) Mountaineering Royalty Disbursement and Operation Guideline. Ministry of Culture, Tourism and Civil Aviation, Government of Nepal. Available at: https://www.tourism.gov.np/publications/5 (accessed 16 November 2020).

MoCTCA (2016a) *Nepal's National Tourism Strategic Plan, 2016–2025*. Ministry of Culture, Tourism and Civil Aviation, Government of Nepal, Kathmandu.

MoCTCA (2016b) *Nepal Tourism Statistics 2015*. Ministry of Culture, Tourism and Civil Aviation, Government of Nepal, Kathmandu.

MoCTCA (2018) *Nepal Tourism Statistics 2017*. Ministry of Culture, Tourism and Civil Aviation, Government of Nepal, Kathmandu.

Murphy, P.E. (1985) *Tourism: A Community Approach*. Methuen, New York.

NPC (n.d.) Periodic Plans. National Planning Commission, Government of Nepal. Available at: https://www.npc.gov.np/en/category/periodic_plans (accessed 22 December 2019).

Nyaupane, G. and Poudel, S. (2011) Linkages among biodiversity, livelihood, and tourism. *Annals of Tourism Research* 38(4), 1344–1366.

Scheyvens, R. (2002) *Tourism for Development: Empowering Communities*. Prentice-Hall, Harlow, UK.

SNV Netherlands Development Organization and University of Hawaii (2007) *A Toolkit for Monitoring and Managing Community-Based Tourism*. SNV Netherlands Development Organization Asia Pro-Poor Sustainable Tourism Network and University of Hawaii, Oahu, Hawaii.

Thapa, K. (2010) Village tourism development & management in Nepal: a case study of Sirubari village. Available at: https://ecoclub.com/education/articles/488-sirubari-village-tourism-nepal (accessed 22 December 2019).

The International Ecotourism Society (TIES) (n.d.) What is ecotourism? Available at: https://ecotourism.org/ties-overview/ (accessed 22 December 2019).

TRPAP (2007) *Lessons Learned: Nepal's Experience Implementing Sustainable Rural Tourism Development Model of Tourism for Rural Poverty Alleviation Programme*. Tourism for Rural Poverty Alleviation Programme, Kathmandu.

UNEP-UNWTO (2005) *Making Tourism More Sustainable – A Guide for Policy Makers*. United Nations Environment Programme and United Nations World Tourism Organization, Madrid.

UNESCO World Heritage Centre (n.d.) World Heritage List. United Nations Educational, Scientific and Cultural Organization. Available at: https://whc.unesco.org/en/list/ (accessed 20 September 2019).

UNWTO (2013) UNWTO Tourism Highlights: 2013 Edition. United Nations World Tourism Organization. Available at: https://www.e-unwto.org/doi/pdf/10.18111/9789284415427 (accessed 14 December 2019).

UNWTO (2018) Tourism and the Sustainable Development Goals – Journey to 2030. United Nations World Tourism Organization. Available at: https://www.e-unwto.org/doi/book/10.18111/9789284419401 (accessed 8 November 2020).

UNWTO (2019) International Tourism Highlights: 2019 Edition. United Nations World Tourism Organization. Available at: https://www.e-unwto.org/doi/pdf/10.18111/9789284421152 (accessed 14 December 2019).

Weaver, D. (2010) Community-based tourism as strategic dead-end. *Tourism Recreation Research* 35(2), 206–208.

WCED (1987) *Our Common Future*. World Commission on Environment and Development. Oxford University Press, New York.

WTTC-UNWTO-EC (1995) *Agenda 21 for the Travel and Tourism Industry: Towards Environmentally Sustainable Development*. World Travel & Tourism Council, United Nations World Tourism Organization and Earth Council, London.

4 Tourism in Afghanistan

Abdul Wassay Najimi*

Aga Khan Trust for Culture, Afghanistan

Introduction

Afghanistan is a landlocked country that has historically been a crossroads of trade routes between South East, North and West Asia (Wikipedia.org, n.d.). It has many historic settlements, villages, cities, places and tourist attractions (Szabo and Barfield, 1991). The tourism industry in Afghanistan developed with government help as it founded the Afghan Tourism Organization, *Garzandoy*, along with Afghan Media in 1958 in Kabul city to promote tourism and tourism hospitality management. The cartographic department produced maps of cities and *Garzandoy* published travel guidebooks to facilitate tourists' visits to the historical cities and archaeological sites in addition to marketing and promotion of tourism (WorldCat.org, n.d.). In the 1960s and 1970s, European tourists travelled in by air and crossed overland through Afghanistan; while the tourists camped at times, the private sector also contributed services in transportation and accommodation. Income from services rendered for the tourists became a visible element in the country's economy. A valid passport with a visa has been required for foreigners to enter Afghanistan. Nationals from Germany and Turkey did not need an entry visa to Afghanistan or could obtain one at arrival, as reciprocal arrangements were in place until 1978/79. Ports of entry were the capital Kabul by air, while border cities like Jalalabad, Herat, Kandahar and Mazar-i Sharif received tourists arriving overland from neighbouring countries like Pakistan, Iran and Uzbekistan (Najimi, 1988; Fig. 4.1).

In 1964 the director of Kabul Museum presented in Venice many photographs of the historic monuments in Afghanistan (Hansen, 2015), while in 1965 Afghanistan became a member of the United Nations World Tourism

*E-mail: aw_najimi@yahoo.com

DOI: 10.1079/9781789246698.0004

Fig. 4.1. Map of Afghanistan, its major cities and border entry points (source: author's own elaboration).

Organization (UNWTO) and invested in some infrastructure to facilitate better inland travel and accommodation of tourists in the Afghan cities and historic sites. The tourism law of Afghanistan (1965), Article One, defined 'tourist' as a person travelling from permanent residence for a while for rest, picnics, holidays, family visit, pilgrimage, medical treatment, study and research as well as missions to another place and returning home. People do travel to other places to see the physical, social and cultural attractions of a place, its people and geography. The moderate Afghanistan and its hospitable people made it possible for tourists to visit tourist attractions at any part of the country and enjoy their stay. Nowadays there are more than 200 places considered as tourist attractions in Afghanistan.

In 1970 revenue from tourism made up a considerable part of the national gross domestic product (GDP) of the country; reportedly about 190,000 tourists visited Afghanistan by land and air arrivals (Hakim, 2010). After the change of the central government in 1978 (a year before the former Soviet Union invaded the country), the tourism law of the country was reformed – the *Garzandoy* was transformed into a government business company, while the Afghan Tour entity was created in 1985. The tourism industry in Afghanistan shrank owing to the civil war and lack of security but also deficiency of the infrastructure needed for tourists to travel around and stay safe. Much of the cultural sites and monuments as attraction sites for tourists got damaged due to war and long conflict, while most of the archaeological sites were illegally excavated. The Kabul Museum, now called the National Museum of Afghanistan, a major place visited by tourists (National Museum of Afghanistan, n.d.), was looted in years 1992–1995 and artefacts were smuggled out of the country. Very few tourists, mainly journalists, visited Afghanistan as the political instability and civil war in the country continued until 2002, when the contemporary regime got established in Afghanistan after decades of war between 1980 and 2001.

The real statistics with respect to tourists visiting Afghanistan are yet to be known but it is also common that people arriving in Afghanistan for jobs try to secure a tourist visa in the first instance. Afghanistan has four international airports, including the Kabul International Airport, Mazar-i Sharif International Airport, Kandahar International Airport and Herat International Airport. Smaller airports serve domestic flights in several provinces of the country. Until 2010, about 175 companies and 80 guesthouses and lodges had obtained operational licences from the government. Some 12 companies operated in sending nearly 10,000 Afghans to Saudi Arabia for the Hajj pilgrimage (Hakim, 2010). According to government sources in 2013, Afghan embassies issued between 15,000 and 20,000 tourist visas annually (Nordland, 2013; Barakzai, 2016). But this figure based on visas issued at Afghan embassies could also include the people and contractors coming for jobs with non-governmental organizations (NGOs) who obtained a tourist visa. One private-sector tourist agency believed that pure tourists are estimated to be only 100 to 150 per year (Nordland, 2013; CEIC, 2020).

Considering local tourism, the statistics on people visiting Bamiyan town in the years 2011 to 2019 indicated an increase from 2500 domestic visitors in 2011 to 13,000 in 2019; and people visiting Band-i Amir Lake increased from about 98,000 in 2014 to 220,000 in 2019. The number of foreign tourists to this province dropped from 1100 in 2011 to 620 in 2019. The government nowadays realizes the role the tourism industry could play in the Afghan economy, but there remain still many challenges facing the industry. The aim of this chapter is to account for the status of development of tourism in the country before the war and the situation today, exploring the policies and promotional tools the government has enacted and the options available ahead.

4.1 Country Background and Tourism Development

The population of Afghanistan is estimated to be about 32.9 million in 1399 (2020/21), of whom around 16.8 million (51.0%) are male and 16.1 million (49.0%) are female. About 23.4 million (71.0%) live in rural areas, 8.0 million (24.3%) live in urban areas and 1.5 million (4.6%) are nomadic populations (National Statistics and Information Authority, 2020). GDP was reported as US$19.9 billion with an estimated GDP per capita of US$630. The inflation rate measured by the overall consumer price index for the country was 0.6% in 2018, whereas it was 5.0% in 2017 (Afghan Government, 2019). Lately, Kabul has had an urban population above 4 million, the city of Bamiyan 14,000, Jalalabad 260,000, Ghazni 66,000, Balkh 550,000, Ghor 8000, Herat 613,000 and Kandahar 491,000 (Trading Economics, n.d.; Afghan Government, 2019) (Fig. 4.1).

Historically, two main trade routes (Silk Road and Spice Road) passed through this land and connected powerful states at lands located in Mesopotamia, Central Asia and Sind, who in history had an eye on this highland and its related skirt-lands for usage of both pastureland and fertile agricultural fields. The rich ancient history, silk trade memories, various cultural communities, beautiful landscapes and natural scenic views of pre-industrial settings of Afghanistan offer a genuine cultural experience to international visitors. The cities that were located along the Silk and Spice Roads were Balkh, Bamiyan, Herat, Jalalabad and Kabul.

Afghanistan has cultural heritage, sites and remains of pre-Islamic and post-Islamic civilizations (Dupree, 2002; Dupaigne, 2007; Aga Khan Trust for Culture, 2018) and is home to over 1500 archaeological sites. Excavations have revealed evidence of both Eastern and Western influences in architecture and material goods (Ball, 1982, 2008). The Zoroastrians, the Greeks, the Kushans and the Sassanians built cities and towns in this part of Asia in the pre-Islamic period (3rd century BC to 7th century AD). The Indus Valley Civilization of the Bronze Age (3300–1900 BC), the Greco-Bactrian (256–125 BC) and Greco-Buddhic civilizations (2nd–1st century BC) and later the Kushan civilization (1st–3rd century CE) are all found at the archaeological sites in Afghanistan. The rulers in the Ghaznavids, Ghurids, Timurids, Safavids and Mogul dynasties contributed to the architecture and urban development through the Islamic periods (7th–18th century AD). Afghanistan has attracted international visitors throughout history: the Chinese Buddhist pilgrim Xuanzang (Hsuan-Tsang), who visited the area on 30 April 630 (Harold, 2006), the Moroccan traveller (geographer) Ibn Battuta in the early 13th century and Alexander Burnes in 1832 (Gibb, 1971).

Places of art-historical or archaeological interest include Ai-Khanoum (possibly the historical Alexandria on the Oxus of the Greco-Bactrian Kingdom from c.280 BCE), Surkh Kotal (a Bactrian site in Baghlan province), the Bamiyan valley (the two monumental statues of Buddha, 37 m and 53 m tall, were carved out of a cliff face overlooking the valley sometime in the 3rd–5th centuries AD), the landscape of Band-i Amir (a series of six water dams created by natural geological formations), Hadda (site of a Greco-Buddhic monastery in Jalalabad province) and the Jam minaret (a 65-m high masonry minaret in a valley in Ghor province, built in the 12th century).

The mountainous land of Afghanistan connects territories in West, South and Central Asia and therefore has attracted travellers for a long time. The land also became a scene of the 'Great Game', the rivalry that occurred between Great Britain and Russia as their spheres of influence in Mughal India, Turkestan and Persia (Iran) moved the two powers closer to one another in south-central Asia from 1830 and lasted throughout the 19th century. After World War II, when tourism became accepted as an industry by mainly Western communities, Afghanistan also became a thoroughfare land for tourists, attracting an increasing number of foreigners, and from the mid-1960s, it became integrated into international tourism. Visitors from Pakistan and India formed the main tourist group at that era. Western tourists were groups that mainly arrived by air, but common also were young Westerners travelling overland to or from India and Nepal (Grötzbach, 1983). Most foreign tourists visited Afghanistan only in transit, but many also visited the main Afghan cities and towns for cultural tourism. International tourism contributed significantly to the country's economy (Bechhoefer, 2002).

The modern airport in Kabul was built with Soviet assistance in the early 1960s while the Afghan Aviation Authority was established. The airport in Kandahar was built with assistance from the USA in 1962. Ariana Afghan Airlines in cooperation with Pan American Airways operated between Kabul, Delhi, Islamabad, Tehran, Tashkent, Istanbul, Jeddah, Frankfurt and London. Planes from India, Tashkent and Islamabad also arrived at Kabul airport. Another airline (Bakhtar Alwatana), for domestic flights with smaller planes, was established in 1967; it kept its name until 1985 when it was renamed Bakhtar Afghan Airlines. The company was then absorbed by Ariana Afghan Airlines and ceased operations in February 1988 (Centreforaviation.com, 2020).

Hotels were built in all major cities of Afghanistan to facilitate the stay of international tourists and visitors. As the author remembers, public hotels used by Western tourists in Kabul city in the 1960s were Kabul Hotel (built in 1945), Ariana Hotel (built in the 1950s) and Spinzar Hotel (built in 1960). Kabul Hotel also had the Ariana Afghan Airlines' ticketing office and a shuttle to the airport. Later the Kabul Inter-Continental Hotel was built on higher ground near the Bagh-i Bala Palace with a good view of the town. This hotel, with its 200 rooms and equipped with a swimming pool, gym and about four restaurants for dining in or room service, served as the nation's first international luxury hotel and was (after Kabul Hotel) one of the most visited by foreigners since its opening in 1969. There were many private hotels that served international tourists; Mustafa Hotel, Park Hotel, Jamil Hotel, Metropole Hotel and the Plaza Hotel were well known in the city centre. Residential buildings were altered to provide smaller guesthouses and casual accommodations around Chicken Street in the new part of Kabul city, where the hippie tourists mostly stayed (Nordland, 2018). Similarly, in other major towns, there were government-run and private-sector-operated hotels. In Herat, the Park Hotel was well known. Later Moafaq Hotel, Behzad Hotel and New Behzad Hotel operated as private-service hotels for the tourists crossing into or out of the border with Iran. In Kandahar, the Ahmad Sha Baba Hotel was well known. Mazar-i Sharif had a hotel run by the government, while other smaller private hotels also operated. In smaller towns and districts, like Maimana, Qadis, and Qala-i Naw of Badghis, smaller lodges, operated under the auspices of the *Garzandoy*, served the tourists. In the eastern town of Jalalabad, the main accommodation was the Spinghar Hotel. Nuristan also had a small guest lodge in Kamu village. Tourist accommodation in Bamiyan was concentrated in the main town and the Bamiyan *Garzandoy* Hotel. There have been plans to develop a hotel complex at Bamiyan since the 1970s (Bechhoefer, 2002).

Providing informed access for tourists to the historical and archaeological sites, the Afghan tourist organization produced travel guidebooks in the 1960s and 1970s and hired English- and French-speaking young travel guides from high schools of Kabul. A high school located in a building adjacent to the Kabul Hotel educated young Afghan students through a three-year programme in the tourism service management and hospitality industry. A survey conducted in 19 hotels in Kabul in 1978/79 revealed the structure, motives and preferences (for sightseeing and shopping) of 734 foreign tourists, shortly before tourism was cut off by political events (Grötzbach, 1983). Then tourists in Afghanistan dropped in numbers and the last organized tour was around 1979. As a result of the Soviet invasion in Afghanistan in 1980, the country went into civil war until the Soviet army withdrew in 1988.

The country did not recover from the political instability for long, as the Mujahideen could not agree on a system of government, Kabul had fierce urban war (1992–1995) and then the Taliban took over from 1995 to 2001. The political atmosphere of those days, particularly during the Kashmir conflict and the demolition of the Babri Masjid, a 16th-century mosque in Ayodhya, India in 1992, in addition to the indigence caused by the international sanctions against their regime, led the more puritan Taliban to seek revenge through the infamous destruction of the Buddha statues of Bamiyan in 2001. This may have been avoidable, arguably, had a concerted effort been made to engage with the more moderate and influential Taliban leaders of the time (Najimi, 2011).

The civil war in Afghanistan had a severe and destructive impact on the physical and social fabric of cities of the country. The regrettable victims in this conflict, however, which were entrapped, have been the historic neighbourhoods in the cities, where fierce street-to-street fighting and use of heavy weaponry resulted in the utter destruction of their fine residential fabric and left some of the valuable historic buildings and monuments in ruin. Like much important urban and architectural heritage in Islamic cities, Kabul evolved through the agent of continued cultural participation, which entrusted deeper meanings to its architecture and urban form. Historically, Kabul, with its wonderful traditional architecture of courtyards, lavish Timurid (later Mughal) gardens, decorated worship centres and its glowing historic bazaars that enticed trade enterprises and merchants around the region, sustained a vibrant city life for its citizens (Kazimee and Najimi, 2017).

4.2 Importance of Tourism for the Economy

Tourism is an important industry and one of the main sources of revenue for many countries in the world. Also, it plays a considerable role within the context of the political relationships between countries and communities, social exchanges and economic development. The Afghan governments have considered tourism an important source of income for the national economy. In the 1970s, tourism in Afghanistan had a major impact on boosting the Afghan economy and providing job opportunities in the private sector. In 1977, 120,000 tourists visited Afghanistan and contributed about US$17 million to the national revenue (Hakim, 2010). The Afghan Tour Company earns profit on issuing operation permits to local firms and tourist services. It charges 110,000 Afs (Afghanis) (about US$1500) for a one-year operation licence to each tourism operator/firm. The revenue from this fee jumped from US$2000 in 2002 to about US$281,000 in 2008 (Hakim, 2010), indicating the considerable tourism development that took place. From 65 guesthouses in 2006 the volume jumped to 430 in 2010; visitors on tourist visas increased from 500 in 2006 to 2000 in 2010. State revenue from the tourism sector was higher in 2009 and declined afterward (Gulistani, 2012). The Afghan Tour company made a profit of 10,278,000 Afs (US$154,500) in 2016, 9,435,000 Afs (US$136,000) in 2017 and 9,700,000 Afs (US$129,000) in 2018 (Afghan Government, 2019, p. 183).

The cultural visitors coming to Afghanistan regenerate the market as they contribute to airplane services, taxi drivers, hotels, tourist guides, handicraft sellers and finally the state, with the tax collected from the services rendered to tourists and income made. The private companies involved in the tourism industry generate a revenue of 1 billion Afs on an annual basis through the services carried out by the sector (TOLOnews, 2016). The Afghan Ministry of Information and Culture (MoIC) indicated in a seminar on tourism in 2017 that Afghanistan generated US$25 million in revenue from the tourism industry in 2016. According to the MoIC, the number of domestic tourists has been higher compared with foreign tourists, who were negligible or almost non-existent. The local tourists have access to about 120 guesthouses across the country, which in turn provides jobs for an estimated 6000 people (Amiri, 2017). More than 1 million pilgrims visit the shrine of Ali in Mazar-i Sharif city for the new year celebration on 21 March every year. Nearly 8000 people visited Kabul's Babar Garden and about 7000 people visited the Qargha Lake in Kabul in 2017. According to statistics of hotel and airline services, more than 400 foreign tourists visited Bamiyan province in June 2019, a significant increase compared with the previous year, with 180 tourists visiting the area in the same season in 2018 (VOA, 2019). Some 44,000 people have been recruited in tourism and 1500 tour and travel agencies acquired licences in the years 2017 to 2019.

Afghans also travel to Uzbekistan and Tajikistan for leisure and trade. Similarly, Afghans travel to Pakistan and India for a visit, trade and medical treatments. Consular sections of embassies of the Western countries seldom work from Kabul, so Afghans must travel to Islamabad or Delhi for paperwork and visas. Afghans also travel to Dubai and Turkey for visits and tourism and many have procured residential properties there.

Some Western tourists visiting Uzbekistan managed to secure their visas and crossed the border bridge over the Amu Darya River into Afghanistan (Dalrymple, 2018). The Couchsurfing app and similar social media sites are also operating in Afghanistan; Kabul alone has about 1000 Couchsurfing hosts. In December 2017, it was estimated that 36% of visitors who had written references for their Kabul hosts on the Couchsurfing site were women. A significant number of the Afghans hosting the Couchsurfing tourists happen to be members of the Taliban. The number of these encounters remains vague because they rarely make it into official records (Dalrymple, 2018). Afghanistan reported a state revenue of about US$486,000 from the tourism sector in 2016 (Panahi, 2017).

International tourism receipts, by definition, are expenditures by international inbound visitors, including payments to national carriers for international transport. The latest value for international tourism receipts in Afghanistan was US$19 million in 2017. Between 2008 and 2017, the value for this indicator ranged from US$179 million in 2013 to US$19 million in 2017 (Fig. 4.2). These receipts include any other prepayment made for goods or services received in the destination country. They also may include receipts from same-day visitors, except when these are important enough to justify separate classification. For some countries, they do not include receipts for passenger transport items (Indexmundi.com, 2019).

4.2.1 Private sector

Tourism has an unbreakable relationship to private-sector services. The companies operating in tourism must obtain a licence from the government. Some

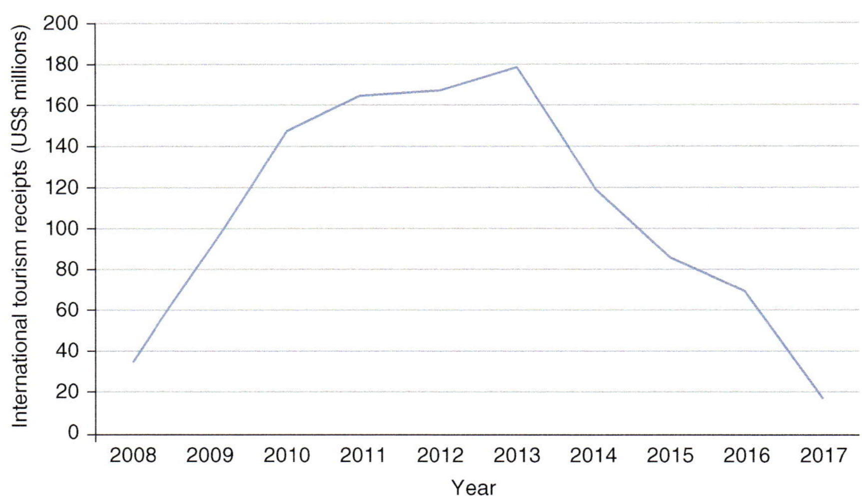

Fig. 4.2. International tourism receipts in Afghanistan, 2008–2017 (source: Indexmundi.com, 2019).

80 guesthouses and lodges as well as 120 restaurants operated at or near the tourist sites to serve local and international visitors until 2010 (Hakim, 2010). According to the Afghan MoIC's tourism department (*Garzandoy*), more than 2000 travel agents were licensed in the country in 2017, but only about 1000 were active in providing tourism services (Hakimi, 2017). Tourist companies could be grouped into three sections: (i) companies having representatives in all provinces and cities and organizing pilgrimages to Iraq (the shrine of Hussain in Karbala), Iran (the shrines) and Saudi Arabia (the Hajj and Umra in Mecca and Medina cities); (ii) companies that only sell tickets through their representative offices; and (iii) tourist companies which bring foreign tourists into Afghanistan (they number about five in Kabul, five in Badakhshan and one in Bamiyan). Tours are organized in groups of different participation and duration depending on the request of the customers or package selection. The longest tour of 14 days offered for international tourists included visits to the cities of Kabul, Bamiyan, Mazar-i Sharif, Herat, Samangan, Kandahar, and the Panjshir valley; here some cultural tours are for walking in the natural landscape. Tourists visiting Badakhshan are mostly foreigners crossing the Amu River from Tajikistan and to explore Ishkashem and Wakhan area and are regulated by Badakhshan tourism companies (Voyakin, n.d.).

Local companies often have collaboration with international tourist companies bringing tourist groups into Afghanistan, like the adventure travel company Hinterland Travel, from the UK, which reported the year 2014 as its busiest year for bookings to Afghanistan (Bridge, 2014). Extreme trips were available to the upscale market, costing around £4950 each for two weeks, excluding flights (Bridge, 2014). Wild Frontiers, another specialist UK adventure travel company, saw an increase in bookings for group tours to Afghanistan. Responsible Travel offered tours taking in the historic quarters in the capital Kabul, mountain villages, lakes, Buddhist sites and ancient forts. But such destinations were limited. Kabul, the Panjshir Valley at the foothills of the Hindu Kush, the caves of Bamiyan, the natural lakes at Band-i Amir, the shrine and the city in Mazar-i Sharif and Balkh, and the city of Herat are easily visited by foreign tourists, mostly travelling by air. Afghan Logistics and Tours, a travel company in Kabul, indicated that more than 100 tourists used their services in 2013; the number reduced to 50 tourists in 2017. In 2016, Wild Frontiers, a British adventure travel company, cancelled its 'Afghan Explorer' tours because of Afghanistan's weak security (Dalrymple, 2018). Charities have also organized tours to Afghanistan. The late Bruce Wannell (died 29 January 2020, age 67) and Henderson Alexander Gall (Sandy Gall, of the Sandy Gall's Afghanistan Appeal (SGAA)) brought organized tours of British seniors to Afghanistan, visiting Kabul, Panjshir, Balkh, Bamiyan and Herat in the years 2002 to 2019.

4.2.2 The role of the government

Usually, the private sector would offer the hospitality services to tourists, but the government also needs to have policies, rules and regulations for the provision and extension of such services and for tourists to have a secured environment for movements. Afghanistan is a member of the UNWTO and would adhere to the Global Code of Ethics for Tourism adopted by the General Assembly of the UNWTO in 1999 (UNWTO, 2001). The government recognizes that Afghanistan has rich tourist places, built cultural heritage and natural landscape suitable for ecotourism. The country, however, lacks good tourism infrastructure and the tourism sector has a shortfall of professional manpower. Opportunities exist whereby domestic and national consortiums could invest in the sector and develop new attractions. Absence of security for travellers and tourists to freely visit places and lack of adequate infrastructure have been the main threats so far (Naylor, 2007). Lately, the US government and the Taliban as confronting parties signed a peace process document on 29 February 2020. This brings hope that the country may return to normality once the Afghan administration and Taliban also make peace. Planning, development and management of tourism projects must be balanced with the local cultural acceptance and possibilities. Both the government regulatory body *Garzandoy* and the private sector involved in the tourism industry would need to market better services and develop tourism along with providing adequate infrastructure for travel and stay in the cities and provincial towns. Development in the tourist industry also calls for educating young people for jobs as guides and functionaries in the hospitality industry, which includes restaurants, hotels, amusement parks and other tourism-related services.

4.3 Afghan Government Policy for Tourism

The first regulatory bill on travel and stay of foreigners in the Kingdom of Afghanistan was issued in 1939 (Table 4.1). The Afghan Tourism Organization, *Garzandoy*, was founded in 1958. The bill was amended in 1965 (*Official Gazette* (*Rasmi-Jareeda*) No. 7, 1344 (1965)) under which the tourism sector of Afghanistan became regulated as services in the tourism industry. The bill was updated in 1975 after the country became a republic. In 1979, the tourist regulatory body *Garzandoy* became part of the Ministry of Transportation. Then, the organization became part of the MoIC after 1992, when the Islamic Government of Afghanistan was established by the Mujahideen. The latest version of the Afghan tourism law is from 2015 (and revised in 2018), which regulates the establishment and operation of the company Afghan Tour and the Afghan Hotel Corporation and mandates the licensing of hotels and travel operation businesses. Although there is no blueprint in place for the development of the tourism sector, in a draft Afghanistan National Ecotourism Strategy (ANES), prepared with and sponsored by the Asian Development Bank in 2005, the government indicated a proposal for the revival of the tourism industry. Understanding clearly that tourism will take time to revive, depending on the security situation, the government wanted to encourage building of the infrastructure by the private sector necessary for tourism to grow, and elaborated the Afghanistan National Development Strategy (ANDS) 2005 (Afghan Government, 2005, p. 86).

Table 4.1. Afghan government tourism policy (source: author's own compilation).

Year	Activity/documents	Authority	Remarks
1939	Regulatory bill on travel and stay of foreigners in Afghanistan	The Kingdom of Afghanistan	Amended in 1965
1958	Afghan Tourism Organization (*Garzandoy*) was founded	The Afghanistan Republic Government	
1965	Amended the 1939 Bill	Democratic Republic of Afghanistan, Ministry of Justice	*Official Gazette* (*Rasmi-Jareeda*) No. 7, 1344 (1965)
1967	Bakhtar domestic airline, Bakhtar Alwatana, was established	Ministry of Transport and Tourism	
1975	Updated the Regulatory Bill of 1965	Afghan Government	
1979	The tourist department, *Garzandoy*, became part of the Ministry of Transport and AIR Services, but more as a business company	Afghan Government, Ministry of Transport	
1985	Bakhtar Alwatana was renamed as Bakhtar Afghan Airlines	Afghan Government, Ministry of Transport	
1988	Bakhtar Afghan Airlines ceased and merged with Ariana Afghan Airlines	Afghan Government, Ministry of Transport	
1992	*Garzandoy* became part of the Ministry of Information and Culture (MoIC)	The Islamic Government of Afghanistan	
1994	Pamir Airways (private sector) founded; flight operations launched in 1995	Ministry of Civil Aviation of the Islamic State of Afghanistan	
2003	KamAir Airline (private sector) was established for domestic and international flights	Islamic Republic of Afghanistan, Ministry of Transport and Civil Aviation	It has an annual turnover of US$200 million and carries more than 1 million passengers each year
2005	Afghanistan National Ecotourism Strategy (ANES), a vision for the revival of the tourism industry, was drafted	Afghan Government and donors	
2006	Dubai- and Kabul-based Safi Airways was established	Private company	
2008	Pamir Airways was taken over by Kabul Bank	Kabul Bank	The licence of the airline was withdrawn, officially due to the poor safety record, in March 2011

Continued

Table 4.1. Continued.

Year	Activity/documents	Authority	Remarks
2015	Afghan tourism law was updated, and Afghan Tour became a company	Afghan Government, Ministry of Culture	
2016	Safi Airline suspended		Afghan authorities forced this company to suspend operations over unpaid debt
2018	Afghan tourism law was revised	Afghan Government, Ministry of Culture	
2020	Plans to revive the Bakhtar airline for domestic flights	Afghan Government	

The draft ANES sets out the rationale for positioning Afghanistan as an ecotourism destination; reviews the policy and institutional context; identifies the potential ecotourism markets, destination and products; and lays out strategies for product development, promotions and communication, and human resource training (Hakimi, 2017). Established in 1958 in Kabul city, the government-owned business entity, Afghan Tour has had to promote tourism and related services in collaboration with the private sector. It has included research and marketing in its policy, working with tourist organizations and UNWTO member countries in attracting international tourists to elevate the sector and help the national economy.

Despite widespread tensions and ongoing wars against the Taliban insurgency, officials in the Afghan government claim that tourism in Afghanistan will return as soon as peace is made. International donor countries have worked with local government leaders on several tourism issues to assist in restoring Afghanistan's natural landscape, rebuild its economy and to bring about a peace deal. The basic policy of the Government of Afghanistan is to promote the development of tourism through the private sector and for a sustainable socio-economic development, protection and valorization of Afghanistan's natural and cultural heritage. A detailed inventory of heritage sites is being undertaken, and the most significant heritage sites will be identified. This is being done in coordination with NGOs already working on the preservation of heritage areas. Conservation plans for a number of heritage sites (rehabilitation and preservation) are drafted and the implementation would help efforts to raise public awareness, provide economic incentives to private owners, and improve tourism development, legal protection, public investment and outreach to international bodies (ANDS 2008–2013) (Afghan Government, 2008, p. 104). Tourism, culture and media can be further developed through effective and fruitful cooperation with regional partners (ANDS 2005) (Afghan Government, 2005, p. 138). Protected area management also will preserve potential ecotourism sites for the future (ANDS 2005) (Afghan Government, 2005, p. 110). President Ashraf Ghani in a seminar on tourism in 2017 reassured the people of Afghanistan about the return of peace and security in the country and wished for the revival of Afghanistan's tourism industry to be used as a resource against terrorism (Amiri, 2017). The government also announced plans to revive the Bakhtar airline in addition to the existing Ariana and KamAir airlines in the country (Khaama Press, 2020) to facilitate more domestic air travel to various places in the country.

4.4 Contemporary Developments since 2001

With the arrival of the Western army to Afghanistan in 2001, the fall of the Taliban regime and the signing of the Bonn Agreement to establish the contemporary regimes, Kabul – the capital city – experienced enormous physical transformation. The built area in the city expanded as some refugees returned from neighbouring countries and the diaspora from the West returned for work and settled in the capital. Housing became a demand. Also, people from provinces moved into Kabul and built homes in the peripheries and hillsides; Kabul became a hub of the activities and home to the army and organizations funded by the West, thus providing jobs, while the rural settlements suffered economically because of uncertainty in security and instability.

The hotels of the past were transformed into new functions or style. Situated between the Presidential Palace and the US Embassy, the Ariana Hotel building was transformed into an office/command centre for Western army use in 2001 (Harooni and Graham-Harrison, 2011). The Kabul Hotel (originally built in 1945) was also transformed after a lease agreement with Kabul Municipality in 2005 and got completely refurbished as a luxury five-star hotel run by the Aga Khan Development Network. The Hotel Inter-Continental Kabul run by a government entity became active. Other private hotels and lodges were made to accommodate the visitors and missions of international organizations. Kabul Star Hotel (built in 2011), Park Palace and Safi Landmark Hotel (built in 2005) are a few examples. Many restaurants with different cuisines, Lebanese, Croatian, Indian, French, etc., served the internationals working in Kabul.

The rehabilitation of Afghanistan's cultural heritage was recognized again as a priority after 2002. The first site in Afghanistan to be added to the list of World Heritage Sites was the 12th-century Minaret of Jam in Ghor province in 2002. Also inscribed on the World Heritage List in 2003 was the Bamiyan valley, two years after the Taliban had destroyed two gigantic Buddha statues. In recognition of the immediate threats facing historic monuments in the country, the World Monuments Watch List included the 9th century Masjid-i-Noh-Gunbad (Haji Pyada Mosque) in Balkh in 2007. On the other hand, the Old City of Herat was identified in 2009 as a city undergoing rapid, uncontrolled 'redevelopment' and thus its inscription is still pending (Najimi, 2011).

Bamiyan became again an attractive destination for tourists, both national and international, for its rich Buddhist sites and the landscape including the lake of Band-i Amir (Fig. 4.3). Being a small province in the highlands it has had better a security situation. Hotels were built to accommodate international visitors to the valley; local-scale hotels serve passengers travelling from remote

Fig. 4.3. Local tourists in Band-i Amir of Bamiyan (source: the author).

places in central Afghanistan and through Bamiyan to Kabul. In Mazar-i Sharif and Herat cities, new hotels were built to accommodate the visiting international missions. The tourist companies compete in providing services to tourists. A company from the north of Afghanistan managed to facilitate the visit of 20 Russian tourists to Bamiyan, Herat, Ghazni and Nangarhar provinces in 2013. It is assumed that the country would attract more international tourists interested in nature and history of this part of the world when Afghanistan resumes a peaceful political environment and better governance. Local tourism would also improve to cities and towns with seasonal climate changes, like people visiting Jalalabad and Kandahar in winter and Kabul, Bamiyan and Badakhshan in the summer holidays.

4.5 Issues and Challenges for Tourism Development

Afghanistan is a war-torn country that has barely been out of the headlines for about four decades. The cultural heritage got damaged and still risks further destruction, a huge potential loss to the people and an obstacle to developing the tourism industry (ANDS 2005) (Afghan Government, 2005). Security is a major issue in the country today, not only for travellers but also for local inhabitants. Understanding that the tourism industry could be important in the Afghan economy, it still faces many challenges. Destinations in Afghanistan attractive for international tourists are of limited access today. Most countries issue a warning against travelling into Afghanistan. The tourist companies admit that insecurity is one of the serious challenges. Despite the wish of some tourists often asking to visit places such as Kapisa, Nuristan and districts of Dara-i Ghorband, the touring companies would not arrange visits to such unsecured places. Tour operators need to seek local advice and consult contacts in certain areas when planning trips and to avoid the military hotspots.

The main rivals in Afghanistan today are the US-backed government and the deposed Taliban. Other armed groups and criminals have a stake in the country's instability. The general rural communities are caught in the conflict and suffer most; as in all conflicts, innocent civilians are killed. Some believe that militant groups such as the Taliban do not necessarily target tourists and the government reports that no foreign tourists are killed, injured or abducted in Afghanistan (Afghan Tourism Organization, n.d.). However, we know that Western people travelling to remote rural places have faced trouble and some have lost their lives. Such an example is the kidnapping of a South Korean tour group on the Ghazni–Kandahar road by the Taliban in 2007. Two of the Koreans were killed and 19 were freed after six weeks in captivity and the payment of a ransom reportedly. Ten members of International Assistance Mission (IAM) aid workers – six Americans, a Briton, a German and four Afghans – were killed in Kuran-wa-Munjan district of Badakhshan province in 2010. A newlywed couple, a pregnant American woman and her Canadian husband, who apparently arrived in Afghanistan in early October and disappeared by mid-October 2012, were reportedly kidnapped in an insurgent area. The couple had embarked on a journey first to Russia, then central Asian countries of Kazakhstan, Tajikistan and Kyrgyzstan, and finally to Afghanistan (Associated Press, 2012). They were travelling partly on foot, staying in a tent and in local guesthouses, until they disappeared. Several secured guesthouses and hotels got attacked between 2008 and 2018 (Nordland, 2013). It is obvious and people wishing to travel to Afghanistan must understand that, in places, tourist infrastructure is almost non-existent. The levels of services will not match those offered in more regular tourist locations (Gordon, 2014).

As Afghanistan tries to rebuild its war-stricken country, and the urge for peace has always been discussed, once peace prevails foreign expertise could be needed to provide advice and knowledge to the tourist industry, as more business travellers are expected over the forecast period. Leisure tourists from neighbouring countries who are used to the insecure environment do not mind travelling into the country to visit friends and relatives and to enjoy the stunning scenery. However, most travellers from places such as the USA, Europe and East Asia would not want to take the risk of entering Afghanistan during the forecast period unless there is a vast improvement in its political development and security (Euromonitor, 2014).

4.5.1 Visas and costs

A visa is required for all non-Afghan nationals. The only exemptions are travellers having other nationalities but born in Afghanistan or born to Afghan parents born in Afghanistan. The visa is issued by the Afghan Missions abroad. Tourists cannot get a visa on arrival. According to the Afghan Travel and Residency Law for Foreign Nationals (Afghan Government, Ministry of Justice, 2015), there are 13 types of visas that are issued by the Afghanistan Missions abroad (Afghan Embassies, n.d.). The visa fee varies according to nationality and type of passport used. For US passport holders, the visa fee is US$160. All US government and NATO contractors obtaining a visa for the first time are required to pay an additional US$200 for a one-year multiple-entry visa (total US$360). Other countries' passport holders obtain a Working Entry Visa for US$100 while they need to pay US$80 for a Tourist Visa or Visit Visa (Afghan Embassies, n.d.). India and Pakistan passport holders get a visa free of charge. Diplomats, officials and United Nations (UN) passport holders also get a visa free of charge. A Tourist Visa is issued for foreign nationals visiting Afghanistan individually or with a group for the purpose of visiting natural, touristic and historic sites or their relatives. The Tourist Visa is valid for 90 days, allowing a period of stay of 30 days, but can be extended only once by the Ministry of Interior with the consent of the Afghan Tourism Department. A Visit Visa is issued for foreign nationals travelling to Afghanistan for participation in sports competitions, exhibitions, visiting relatives and friends, and conducting a pilgrimage (people from India and other Muslim countries) to holy sites in Afghanistan. A Transit Visa is issued to foreign nationals with a valid visa crossing overland through Afghanistan to a final destination (Afghan Embassies, n.d.). Certain embassies ask for a letter of invitation, while others may not. Some embassies do not issue tourist visas due to safety concerns (Caravanistan.com, n.d.).

The cost of living in Afghanistan has become expensive. In 1975, a regular meal in an Afghan restaurant would cost about 25 Afs (about US$1); maybe a bit more in tourist hotels. In 1999, the UN estimated the daily cost of staying in Kabul at US$70. Today, of course, the cost of living is higher. A simple meal in a roadside restaurant would cost 150 Afs. The hotels accepting foreigners charge international rates. Serena Hotel might charge US$250–400 for a room; Kabul Star Hotel charges US$350; Park Star Hotel charges US$180 (Booking.com, 2020a). In Bamiyan, the Silk Road Hotel charges US$80 for a single room and US$100 for a twin room (Silk Road Hotel, n.d.). In Herat, a hotel may cost about US$100–150 (Booking.com, 2020b). The bus fare from Kabul to Herat is 1000 Afs, and that from Kabul to Mazar-i Sharif about 600 Afs. A Kabul–Herat air ticket costs about US$100. Another issue is that there is no direct flight to European destinations. Earlier, Safi Airways attempted to serve Kabul–Frankfurt flights for a while between 2009 and 2010 but was stopped due to an EU ban barring all Afghan carriers from flying into Europe for lack of technical qualification. Emirates Airlines operates through Dubai since 2013. Turkish Airlines serves flights through Istanbul. Ariana Afghan Airlines and KamAir also fly to Istanbul and some capitals in the region.

4.5.2 Fee to visit historic sites

Normally there is no fee to enter a historic site in Afghanistan. To manage sustainable resources for administration and maintenance of the sites, entry tickets have been introduced at some sites. Entry ticket to the restored historic garden of Babur and Chihilsutoon Garden is 30 Afs for adult nationals (US$5 international visitors) (S. Arash, 2020, Chihilsutoon Garden's activity report 2019, unpublished; A.S. Wardak, 2020, Babur Garden's activity report 2019, unpublished). The ticket for day visits to the heritage sites of Bamiyan is 60 Afs (less than US$1) per local person and 300 Afs (less than US$5) for foreign visitors. An entry ticket for the Band-i Amir natural site is issued by the vehicle, depending on the capacity and number of its passengers, and varies from 100 to 450 Afs.

Conclusions

Over the course of history, a rich diversity of peoples has come and dwelled in this land that, as a hub, let great civilizations meet. The landscape and the cultural heritage in Afghanistan are rich attractions to travellers, researchers and tourists. While tourism in this country gained momentum in the 1970s, it declined because of political conflicts since the 1980s. After a long period of conflicts in Afghanistan, now the Afghan government has an opportunity to rebuild the nation and develop the tourism industry once the political environment is peaceful through the dialogue for peace that the US government and Taliban have been conducting. Income from the tourism industry could be a considerable source of revenue for the national finance and local job markets. The Afghan government today is more focused on an open market economy, therefore once peace is secured the potential for the private sector to invest in the tourism industry and required infrastructure is large. The tourism industry would bring changes in the life of the people, and the regulatory bodies should make sure that culturally sensitive aspects are preserved and protected from unwanted influence (Atiqzada, 2019) and the Global Code of Ethics for Tourism is respected. Foreign expertise and investment could be needed in some context to expand the services and infrastructure needed for the hospitality of international tourists to Afghanistan; this would create direct or indirect employment in the tourism sector.

The findings reported in this chapter must be seen in light of some limitations, such as lack of previous research studies on the topic as well as insufficient statistical information and limited access to data in different organizations in a country heavily broken by war and instability over the years. Continued research is needed with a more gradual and focused approach, in the short and medium terms, to chart out opportunities and improve security, infrastructure and services for ecotourism in Afghanistan.

This country has a rich cultural heritage and natural beauty and has the potential to attract once again a significant number of tourists, not only from South Asia (ANDS 2005) (Afghan Government, 2005, p. 86), but also cultural tourists from other parts of the world. Tourism can play a role in national economic growth, as it attracts investments that generate jobs and services that can be part of the development of the country.

Acknowledgements

The author thanks architect Zaher Shewa, Head of the Department of Urban Planning, Bamiyan province, for helping with some references on statistics; architecture student Ms Hadya Waqfi, finding monographs used as references for this research; and the editors and the anonymous reviewers for their comments on the manuscript.

References

Afghan Embassies (n.d.) Consular Office, The Embassy of Afghanistan in Washington, DC. Available at: http://www.afghanembassy.us (accessed 22 October 2019). Consular Office, The Embassy of Afghanistan in London. Available at http://www.afghanistanembassy.org.uk (accessed 28 February 2020). Consular Office, Afghan Ministry of Foreign Affairs. Available at: https://www.econsulate.gov.af (accessed 28 February 2020).

Afghan Government (2005) Afghanistan National Development Strategy 2005: An Interim Strategy for Security, Governance, Economic Growth and Poverty Reduction. Available at: https://my.nps.edu/documents/105988371/107571254/National+Development+Strategy.pdf/8176334f-927b-495a-81e2-d899c57c1afb (accessed 28 February 2020).

Afghan Government (2008) Afghanistan National Development Strategy 1387–1391 (2008–2013): A Strategy for Security, Governance, Economic Growth and Poverty Reduction. Available at: https://www.wto.org/english/thewto_e/acc_e/afg_e/WTACCAFG18_CD_1.pdf (accessed 9 November 2020).

Afghan Government, Ministry of Justice (2015) The Travel and Residency Law for Foreign Nationals in Afghanistan. *Official Gazette* no. 1185, 11/7/1394 (3 October 2015). Available at: https://moj.gov.af/en/official-gazette (accessed 28 February 2020).

Afghan Government (2019) *NSIA Yearbook 2018–19*. National Statistics and Information Authority, Kabul.

Afghan Tourism Organization (n.d.) History of Afghan Tourism Organization. Available at: http://tourism.gov.af (accessed 23 July 2019).

Aga Khan Trust for Culture (2018) Restoration in Kabul. Available at: https://www.akdn.org/where-we-work/central-asia/afghanistan/cultural-development/restoration-kabul (accessed 28 February 2020).

Amiri, S (2017) Ghani calls for the revival of Afghanistan's tourism industry. *TOLOnews*, 23 July 2017. Available at: https://tolonews.com/afghanistan/ghani-calls-revival-afghanistans-tourism-industry (accessed 8 March 2020).

Associated Press (2012) Family pleas for US woman, Canadian husband missing in Afghanistan. *CP24 News*, 30 December 2012. Available at: https://www.cp24.com/news/family-pleas-for-u-s-woman-canadian-husband-missing-in-afghanistan-1.1095949 (accessed 3 June 2020).

Atiqzada, R. (2019) Building opportunities for growth and job creation in Afghanistan. Presented at the *13th UNWTO Asia/Pacific Executive Training Program on Tourism Policy and Strategy, 25–28 June 2019, Jeju, Republic of Korea*.

Ball, W. (1982) *Archaeological Gazetteer of Afghanistan* (2 vols). Editions Recherche sur les Civilisations, Paris.

Ball, W. (2008) *The Monuments of Afghanistan: History, Archaeology and Architecture*. I.B. Tauris & Co. Ltd, London.

Barakzai, N.A. (2016) 20,000 foreign tourists visit Afghanistan annually. *Pajhwok Afghan News*, 27 September 2016. Available at: https://www.pajhwok.com/en/2016/09/27/20000-foreign-tourists-visit-afghanistan-annually (accessed 15 May 2017).

Bechhoefer, W. (2002) Afghanistan revisited: sustainable development and eco-tourism. *Traditional Dwellings and Settlements Review* 14(1), 40. Available at: http://www.jstor.org/stable/41757957 (accessed 26 February 2021).

Booking.com (2020a) Park Star Hotel (Hotel), Kabul (Afghanistan) deals. Available at: https://www.booking.com/hotel/af/park-star-kabul.en-gb.html (accessed 16 November 2020).

Booking.com (2020b) Arg Hotel (Hotel), Herat (Afghanistan) deals. Available at: https://www.booking.com/hotel/af/arg.html (accessed 16 November 2020).

Bridge, M. (2014) Wish you were here? Afghanistan attracts adventure tourists. *The Times*, 29 December 2014. Available at: https://www.thetimes.co.uk/article/wish-you-were-here-afghanistan-attracts-adventure-tourists-9jmdk9r36sl (accessed 16 October 2019).

Caravanistan.com (n.d.) The Silk Road Travel Guide. Afghanistan visa. Available at: https://caravanistan.com/visa/afghanistan/ (accessed 28 February 2020).

CEIC (2020) Afghanistan Tourism Statistics, Yearbooks 2008–2018. Available at: https://www.ceicdata.com/en/afghanistan/tourism-statistics (accessed 8 March 2020).

Centreforaviation.com (2020) Bakhtar Afghan Airlines. Available at: https://centreforaviation.com/data/profiles/airlines/bakhtar-afghan-airlines (accessed 26 May 2020).

Dalrymple, S. (2018) Couchsurfing with the Taliban. *The New York Times*, 8 June 2018. Available at: https://www.nytimes.com/2018/06/08/opinion/couchsurfing-with-the-taliban.html (accessed 28 February 2020).

Dupaigne, B. (2007) *Afghanistan: Monuments Millénaires*. Actes sud, Paris.

Dupree, N.H. (2002) Cultural heritage and national identity in Afghanistan. *Third World Quarterly* 23(5), 977–989. Available at: https://www.jstor.org/stable/3993399 (accessed 26 February 2021).

Euromonitor (2014) Travel and Tourism in Afghanistan. Available at: https://www.euromonitor.com/travel-and-tourism-in-afghanistan/report (accessed 22 July 2019).

Gibb, H.A.R. (trans. and ed.) (1971) *The Travels of ibn Baṭṭūṭa, A.D. 1325–1354 (Volume 3)*. Hakluyt Society, London.

Gordon, S. (2014) Bookings for adventure trips to Afghanistan DOUBLE as curious travellers defy Foreign Office advice and pay £5,000 each to visit troubled country. *Mail Online*, 29 December 2014. Available at: https://www.dailymail.co.uk/travel/travel_news/article-2889853/Bookings-adventure-trips-Afghanistan-DOUBLE-curious-travellers-defy-Foreign-Office-advice-pay-5-000-visit-troubled-country.html (accessed 8 March 2020).

Grötzbach, E. (1983) Der Ausländertourismus in Afghanistan bis 1979: Entwicklung, Struktur und räumliche Problematik [International tourism in Afghanistan until 1979]. *Erdkunde* 37(2), pp. 146–159. Available at: https://www.jstor.org/stable/25644639 (accessed 26 February 2021).

Gulistani, A.R. (2012) درآمد توریزم و تاثیر آن در افزایش عاید ملی افغانستان در سالهای 1385 الی 1390 [Tourism Income and Its Impact on Increasing Afghanistan's National Income from 2006 to 2011]. Monograph, Faculty of Economics, Kabul University, Kabul.

Hakim, A. (2010) تحلیل و ارزیابی خدمات هوانوردی ملکی و توریزم در افغانستان [Analysis and Evaluation of Civil Aviation and Tourism Services in Afghanistan]. Monograph, Department of Management and Business Administration, Kabul University, Kabul.

Hakimi, D.A. (2017) Islamic Republic of Afghanistan. Presented at the *11th UNWTO Asia/Pacific Executive Training Program on Tourism Policy and Strategy, 20–23 March 2017, Port Moresby, Papua New Guinea*. Available at: https://webunwto.s3-eu-west-1.amazonaws.com/imported_images/47190/presentation_afghanistan.pdf (accessed 24 February 2020).

Hansen, E. (2015) *The Ghurid Portal of the Friday Mosque of Herat, Afghanistan*. Aarhus University Press, Aarhus, Denmark.

Harold, F. (2006) Bamiyan and Buddhist Afghanistan. Available at: http://depts.washington.edu/silkroad/cities/afghanistan/bamiyan.html (accessed 15 October 2019).

Harooni, M. and Graham-Harrison, E. (2011) Afghan employee kills US citizen at Kabul CIA base (former Ariana Hotel). *Reuters*, 26 September 2011. Available at: https://www.reuters.com/article/us-afghanistan-usa-attack/afghan-employee-kills-u-s-citizen-at-kabul-cia-base-idUSTRE78P0BC20110926 (accessed 8 March 2020).

Indexmundi.com (2019) Afghanistan – International tourism, receipts (current US$). Available at: https://www.indexmundi.com/facts/afghanistan/indicator/ST.INT.RCPT.CD (accessed 8 March 2020).

Kazimee B.A. and Najimi, A.W (2017) An analysis of urban regeneration and architectural heritage saving in Kabul since 2001. *International Journal of Heritage Architecture* 1(4), 671–682.

Khaama Press (2020) Bakhtar Afghan airlines will restart operations with domestic flights. *The Khaama Press News Agency*, 16 January 2020. Available at: https://www.khaama.com/bakhtar-afghan-airlines-will-restart-services-with-domestic-flights-98790875/ (accessed 6 March 2020).

Najimi, A.W. (1988) *Herat: The Islamic City – A Study in Urban Conservation.* Scandinavian Institute of Asian Studies Occasional Papers. Curzon Press, London.

Najimi, A.W. (2011) Built heritage in Afghanistan: threats, challenges and conservation. *International Journal of Environmental Studies* 68(3), 343–361. Available at: https://doi.org/10.1080/00207233.2011.573961 (accessed 26 February 2021).

National Museum of Afghanistan (n.d.) National Museum of Afghanistan. Available at: https://www.nationalmuseum.af/ (accessed 8 March 2020).

National Statistics and Information Authority (2020) Estimated Population of Afghanistan 2020–21. Available at: https://www.nsia.gov.af:8080/wp-content/uploads/2020/06/%D8%A8%D8%B1%D8%A2%D9%88%D8%B1%D8%AF-%D9%86%D9%81%D9%88%D8%B3-%DA%A9%D8%B4%D9%88%D8%B1-%DB%B1%DB%B3%DB%B9%DB%B9-%D9%86%D8%B3%D8%AE%DB%80-%D8%A7%D9%88%D9%84.pdf (accessed 2 June 2020).

Naylor, S. (2007) *Not a Good Day to Die: The Untold Story of Operation Anaconda*. Berkley Books, New York.

Nordland, R. (2013) Despite Taliban threat, Afghanistan manages to entice some tourists. *The New York Times*, 12 January 2013. Available at: https://www.nytimes.com/2013/01/13/world/asia/taliban-targets-tourists-but-some-still-visit-afghanistan.html (accessed 19 February 2020).

Nordland, R. (2018) There are no chickens on Chicken Street. Now there are no customers either. *The New York Times*, 22 April 2018. Available at: https://www.nytimes.com/2018/04/22/world/asia/afghanistan-kabul-chicken-street.html (accessed 18 October 2019).

Panahi, J. (2017) درافغانستان-امن؛-درصد-بودجه-دولت-از-سیاحت-تامین-میشود (In safe Afghanistan 50% of the government budget is provided by tourism). *Afghan News*, 1395/10/30

(19 January 2017). Available at: http://afghannews.af/%D8%AF%D8%B1-%D8%A7%D9%81%D8%BA%D8%A7%D9%86%D8%B3%D8%AA%D8%A7%D9%86-%D8%A7%D9%85%D9%86%D8%9B-%D8%AF%D8%B1%D8%B5%D8%AF-%D8%A8%D9%88%D8%AF%D8%AC%D9%87-%D8%AF%D9%88%D9%84%D8%AA-%D8%A7%D8%B2-%D8%B3%DB%8C%D8%A7%D8%AD%D8%AA-%D8%AA%D8%A7%D9%85%DB%8C%D9%86-%D9%85%DB%8C%D8%B4%D9%88%D8%AF (accessed 22 October 2019).

Silk Road Hotel (n.d.) Hotel Silk Road Bamiyan. Available at: http://silkroadbamiyan.com/hotel (accessed 22 October 2019).

Szabo, A. and Barfield, T.J. (1991) *Afghanistan: An Atlas of Indigenous Domestic Architecture*. University of Texas Press, Austin, Texas. Available at: https://books.google.com.af/books/about/Afghanistan.html?id=IdeBjIR5678C&redir_esc=y (accessed 16 November 2020).

TOLOnews (2016) Afghanistan's tourism generates 1bn Afs a year. *TOLOnews*, 27 September 2016. Available at: https://tolonews.com/afghanistan/afghanistans-tourism-generates-1bn-afs-year (accessed: 8 March 2020).

Trading Economics (n.d.) Afghanistan Population. Available at: https://tradingeconomics.com/afghanistan/population (accessed 3 June 2020).

UNWTO (2001) Global Code of Ethics for Tourism. United Nations World Tourism Organization. Available at: https://webunwto.s3.eu-west-1.amazonaws.com/imported_images/37802/gcetbrochureglobalcodeen.pdf (accessed 1 June 2020).

VOA (2019) سیاحت در افغانستان در حال برگشت است (Tourism return to Afghanistan). Available at: https://www.darivoa.com/a/a-48-2009-07-08-voa3-94152689/1432788.html (accessed: 8 March 2020).

Voyakin, D. (n.d.) The Great Silk Roads. Blog, Pamir Highway tours. Available at: http://pamirhighwayadventure.com/blog/thegreatsilkroad/ and https://en.unesco.org/silkroad/content/great-silk-roads (accessed 26 February 2020).

Wikipedia.org (n.d.) Transasia trade routes 1st C CE. Available at: https://commons.wikimedia.org/wiki/File:Transasia_trade_routes_1stC_CE_gr2.png (accessed 9 March 2020).

WorldCat.org (n.d.) Afghan Tourist Organization. Available at: http://www.worldcat.org/identities/lccn-n79-65472/ (accessed 11 March 2020).

5 Planning for Sustainable Tourism Development in Bhutan

Stephen Pratt[1]* and Wantanee Suntikul[2]

[1]*The University of the South Pacific, Fiji;* [2]*Independent Researcher*

Introduction

The Kingdom of Bhutan is well known for its distinct national and cultural identity based on Vajrayana Buddhism (Shrotryia, 2006). Many elements of governmental policy, such as the principle of Gross National Happiness and a cautious approach to tourism development, are rooted in Buddhist values (Ura *et al.*, 2012). In the quest to preserve its culture and safeguard against some of the negative effects of tourism, policy makers in Bhutan have tightly controlled the way tourism operates in the country (Nyaupane and Timothy, 2010). Tourism is essentially a people-to-people business, and the success of the tourism experience and the impacts on the host community depend on the host–guest relationship (Smith, 2012). This chapter seeks to investigate the ways in which tourism planning and policies in Bhutan promote, reinforce and constrain sustainability. The scope of this work will cover food and beverage purchasing decisions, human resources, transportation and souvenir sales in the tourism and hospitality industry in Bhutan. To achieve these research objectives, 19 in-depth interviews were conducted with hotel general managers in the main tourism areas of Thimphu and Paro. Our findings reveal that, while the controlled nature of tourism in Bhutan certainly protects its residents from the negative excesses of global tourism, numerous policies also impede tourism and hospitality from spreading their benefits more widely. Policy makers in Bhutan would rather adhere to the precautionary principle than allow detrimental practices or risk irreversible impact on Bhutanese culture.

*E-mail: stephen.pratt@usp.ac.fj

DOI: 10.1079/9781789246698.0005

5.1 Tourism in Bhutan

Sustainable tourism is a commonly cited goal of tourism destinations and individual accommodation establishments (Liu, 2003). While sustainable tourism is a universal goal, what sustainable tourism looks like in practice and how it can be achieved are subject to much debate and thinking (Bramwell and Lane, 1993). Sustainable tourism is often categorized into three areas or dimensions – economic, environmental and socio-cultural – although these categories often overlap in discussions on the impacts of tourism. These three dimensions are often used to categorize positive and negative impacts from tourism. Yet these impacts vary in degrees and are often contradictory across and within these three dimensions, as in the case of Bhutan (Brunet *et al.*, 2001).

Bhutan is a landlocked country sandwiched between two powerful neighbours, China and India, with a population of around 750,000. Vajrayana Buddhism is the state religion and the principles of Buddhist philosophy profoundly influence governance and policy in Bhutan (Sharma and Zam, 2019). The Gross National Happiness policy, which embraces happiness, not economic production or material wealth as the indicator of national well-being and development goals, is the best-known manifestation of this perspective (Bates, 2009). Affirming this commitment, Bhutan's Minister for Economic Affairs, Lyonpo Khandu Wangchuk, stated in 2005:

> Our tourism policy is inspired by the philosophy of Gross National Happiness, and emphasizes the harmony between tradition and modernity, between Man and Nature, and the importance of preserving our living local systems in an era of globalization.
>
> (Royal Government of Bhutan, 2005)

As Bhutan is a small, mountainous country with a lack of good-quality tourism infrastructure, the Bhutanese government has taken a cautious approach to planning and developing tourism (Gurung and Seeland, 2008). Suntikul (2018) notes that Bhutan's tourism policy seeks to encompass the government's protectionist philosophy while at the same time charting a path to modernization and economic development. Even though the Bhutanese government recognizes the value of the economic income from tourism, it is also aware of the potential negative impacts of tourism (Dorji, 2001). Uncontrolled development of tourism could also overstrain the country's sparse infrastructure and could bring negative impacts on the environment and social fabric. For this reason, the government's long-standing approach to tourism is described as 'high value, low volume', aiming for a limited number of high-spending tourists as opposed to the mass tourism market.

The history of international tourism to Bhutan dates back to 1974, the year of the coronation of His Majesty the Fourth King Jigme Singye Wangchuk (Ritchie, 2008). Tourists arriving in that year entered Bhutan overland through India as 'paying guests', with each tourist paying a flat daily tariff of US$130, which covered all accommodation, internal transport, meals, attraction fees and tour guides. In 1983, Bhutan opened its first and only international airport. The national air carrier Druk Air was founded in 1988. In 1989, the government increased the tourist daily tariff to US$200, and in 2012 the tariff was again increased, to US$250 during the high seasons and US$200 during low seasons (Suntikul and Dorji, 2016).

Tourist movement is still restricted by a government-decreed list of places that tourists are not allowed to visit, mostly important religious sites. Most tourist traffic occurs in the high seasons of March to May and September to November. This pronounced seasonality (nearly half of bed nights occur in April and October) (Rinzin *et al.*, 2007, p. 115) leads to intense concentrations of tourists at certain times and underutilized tourism infrastructure at other times, although there are currently government initiatives to attract more tourists in the off-season months as well.

Tourism is becoming increasingly important to Bhutan. There are two distinct tourism markets: the international market and regional tourists. These two markets are subject to different policy and planning rules. For regional tourists, countries whose nationals do not need to pay a daily tariff include India, Bangladesh and the Maldives. Diplomatic service and official passport holders from Thailand and Switzerland are also exempted. The number of international tourists to Bhutan grew from 7559 in 2000 to 27,196 in 2010, when the total number of tourists was 40,873 (no statistics were recorded for regional tourists prior to 2010). Table 5.1 shows a continuing upward trend in visitors in subsequent years. In 2013 and 2014, there was a relatively even split between international visitors and regional visitors, but in 2015 there was a near doubling of regional visitors and a

Table 5.1. Total visitors to Bhutan, 2000–2018 (source: authors' own compilation based on data from Tourism Council of Bhutan, various).

	Visitors			Proportion of total visitors (%)		Annual growth rate (%)	
Year	International	Regional[a]	Total	International	Regional	International	Regional
2000	7,559						
2001	6,393						
2002	5,599						
2003	6,261						
2004	9,249						
2005	13,626						
2006	17,344						
2007	21,094						
2008	27,636						
2009	23,480						
2010	27,196	13,677	40,873	66.5	33.5		
2013	52,783	63,426	116,209	45.4	54.6		
2014	68,081	65,399	133,480	51.0	49.0	29.0	3.1
2015	57,537	97,584	155,121	37.1	62.9	−15.5	49.2
2016	62,773	146,797	209,570	30.0	70.0	9.1	50.4
2017	71,417	183,287	254,704	28.0	72.0	13.8	24.9
2018	71,807	202,290	274,097	26.2	73.8	0.5	10.4

[a]Prior to 2010, statistics on total visitors did not include regional tourists.

decrease in international visitors, mainly attributable to the major earthquake in April 2015 adversely affecting international arrivals. While international visitor arrivals started to recover in 2016, the number of regional visitors increased again by 50% and continued its upward trajectory so that by 2018, of the 274,097 visitors, almost three-quarters (73.8%) were regional visitors.

The travel behaviours of the two types of visitors differ in several ways. International visitors generally fly in, while regional tourists enter Bhutan via the land border with India. A desire to experience the nation's unique culture, including festivals, is the strongest driver of visitors to Bhutan. Of Bhutan's traditional festivals, the Thimphu and Paro Tshechus receive the largest number of visitors, with 20% and 32% of all tourists witnessing these festivals, respectively (Tourism Council of Bhutan, 2019). The highly seasonal pattern of tourism to Bhutan, with the highest visitation being in May and October, coinciding with the festivals. The majority of visitors to Bhutan take part in some form of cultural activity, with nature-based and recreational activities also being popular.

5.2 Methodology

To address the research objectives of this chapter, the authors conducted in-depth interviews with general managers of 19 hotels in Bhutan. Both authors travelled to Bhutan to conduct the interviews over two-week periods in 2015 and 2016, with one week in Paro, the location of Bhutan's international airport, and another week in the capital city of Thimphu on each occasion. Hotels in Thimphu and Paro were targeted as these two cities account for 59.4% of total bed nights in the country (Tourism Council of Bhutan, 2019). Purposeful sampling was undertaken by targeting these hotels. Introductions to the hotel general managers were facilitated through the Royal Institute for Tourism and Hospitality, with which the researchers were professionally networked. Snowballing sampling was undertaken after the initial interviews to expand the sample size. Although in absolute terms 19 hotels is not a large sample, it represents over 26% (19 out of 73) of the three-, four- and five-star hotels in Thimphu and Paro. International tourists are only allowed to stay in three-star and above hotels. Regional tourists (from India, Bangladesh and Maldives) are allowed to stay in homestays or hotels.

Of the 19 hotels surveyed, ten hotels were three-star hotels, five hotels were four-star, and four hotels were categorized as five-star establishments. Hotels in Bhutan are generally small by international comparison. The minimum number of rooms of a hotel in the sample was 8, with a maximum of 78. In this sample, the mean number of rooms is 33, containing 55 beds on average. The average length of stay for the tourists' whole trip is 6.63 nights, and given that tourists visit several locations while in the country, the average length of stay at any one hotel is short (Tourism Council of Bhutan, 2019). The average length of stay for the 19 interviewed accommodation establishments is 2.74 nights. The room rates vary depending on the time of year. The research participants reported that the average room rate was approximately US$150 per person per night. Rates,

Table 5.2. Sample accommodation statistics (source: the authors' own compilation).

	Mean	Median
Number of rooms	33	30
Number of beds	55	54
Length of stay (nights)	2.74	2
Room rate (US$/night)	150.31	81.95

however, ranged from US$23.42 for one three-star hotel to US$800 for a luxury resort. Table 5.2 shows basic accommodation statistics from the sample.

Hotel general managers were perceived to be the most appropriate research respondents as they are ultimately responsible for, and had the most knowledge about, the procurement practices, human resources and other services their guests required, such as transportation and shopping. In instances where hotel general managers did not know an answer to a particular inquiry, they either sought the answer from other staff members while the researchers were still on the premises or emailed later with a response. A discussion guide was developed for the semi-structured interviews that captured quantitative data across four areas: food and beverage procurement practices, human resources, transportation and handicrafts/souvenirs. For each of these areas, participants were asked the extent to which these resources were sourced locally and the extent to which they were sourced from abroad. After capturing these data, participants were then asked to provide details on each area as to how sustainable each area was and how the hotel's decisions fit with the ethos of Bhutanese culture. Each interview lasted approximately one hour. The data were then entered from the paper questionnaire into Microsoft Excel for further analysis. The analysis consisted predominantly of descriptive statistics while the qualitative data were manually coded to extract themes.

5.3 Findings

We assessed tourism policy and planning in Bhutan across four areas to highlight the ways that the hospitality sector engages with the planning and policy of the Bhutanese government across four areas: food sourcing, human resources, handicrafts and transportation.

5.3.1 Food sourcing

The sourcing of food relates to the economic sustainability of food systems (Pratt *et al.*, 2018). This entails many aspects. The further food travels, the more energy is used in transportation and storage. This contributes to greenhouse gas emissions. Sourcing food locally links the local agricultural sector and the hospitality sector by strengthening the tourism supply chain (Meyer, 2007). An added benefit is that economic leakages are reduced. Strengthening the supply chain can also help alleviate poverty, specifically for low-income farmers (Hummel and van der Duim, 2012). This has been approached in the past through both supply-side and demand-side approaches. Significant work has been undertaken linking the agriculture and tourism sectors by helping farmers understand the needs of hotels and other food and beverage establishments that cater to tourists (Mitchell, 2013; Pillay and Rogerson, 2013). From the demand side, decision makers concerning menu and food choices, such as chefs, are being educated and encouraged to use more local ingredients in their food and beverage offerings (Bélisle, 1983; Pratt *et al.*, 2017).

Interviews with the hotel general managers revealed that food and beverages was the largest expense area for their operations. The highest food expense is meat. Rinzin *et al.* (2007) also noted this in their study in Bhutan; meat, poultry and fish comprise 33.9% of total food and beverage costs. Overall, 65.5% of all foods and beverages are imported. In general, the higher-starred hotels and resorts import a larger amount of foods and beverages than their lower-starred counterparts. The import quotient is higher for meat, at 71.2%. Most of the local animal products are chicken eggs and river fish. The main obstacles to using more locally produced foods and beverages are threefold. First, as noted above, tourism in Bhutan is seasonal; there is high tourist demand around festival times and very low demand at other times. This means that local farms do not have the quantity of produce to meet demands at certain times of the year. Second, given the limited supply of produce, farmers can charge higher prices. This is a deterrent to hotels that may be able to purchase larger, more consistent quantities at lower prices from nearby India. This was more of a factor for three-starred hotels, who were more budget-conscious than the higher-starred hotels and resorts. Consistency was the other constraint mentioned by interviewees as to why hotels do not use more local ingredients. While it was acknowledged that locally grown food was generally of higher quality, there was more variability in the quality, and it was perceived as more convenient and reliable to purchase foods and beverages through an import wholesaler.

As highlighted by Pratt *et al.* (2018), the highly controlled nature of Bhutan's tourism industry is a limiting factor in linking different sectors of the economy: in this case, agriculture and hospitality. The degree to which tourists can taste or even try local foods is dependent on choices made by tour operators. Tourists do not have discretion over which restaurants they visit; they are taken to eating places by their guides. Some tour guides might take tourists to eating places that offer more locally grown foods while others will not. Therefore, the strengthening of agriculture–tourism links is somewhat constrained by the nature of tourism in Bhutan, in which tourists have little or no discretion as to the food establishments they patronize.

The Bhutanese government has adopted the stance that tourism development should not compromise the values that underlie the protection of cultural values and the enhancement of quality of life. These cultural and religious values also affect food choices. Because Bhutan is a Buddhist country, the slaughtering of animals is considered sinful, although many people consume meat from animals slaughtered by others. Some tourists also prefer meat in their diet, adding to the demand for meat in the country. Indeed, according to

Bhutan's Ministry of Agriculture and Forests' Livestock Department, Bhutan is the highest per capita meat consumer in South Asia, with about 13.5 kg of meat consumed per person annually. Consequently, the bulk of the meat consumed in the country is imported. In 2014, Bhutan imported 10,336 t of meat products worth US$19.2 million from India and Thailand (Dema, 2015). 'Killing is against our religion, but consuming meat has always been a part of culture and tradition', said a government official (Dema, 2015). After some debate about whether the Bhutanese government should open a slaughterhouse, the government reaffirmed its stance not to do so, meaning that meat would continue to be imported. This is an example of tourism planning having both positive and negative effects on different elements of tourism.

5.3.2 Human resources

Most tourism products are affected by seasonality (Goeldner and Ritchie, 2012) and Bhutan is no exception; the peak periods correspond to the major festivals in Bhutan as tourists are motivated to witness Bhutan's unique cultural events. Seasonality can have implications for the sustainability of tourism and the well-being of local employees. In typical tourism markets, during peak periods, staff may be expected to work overtime for no extra money and without adequate breaks (Swarbrooke, 1999). Conversely, in periods of weak demand, staff may be moved from permanent to casual employment or given limited hours, making it difficult for them to develop a career or be guaranteed a regular income.

However, in Bhutan, seasonality for human resources is not a problem, at least in terms of employing staff (Table 5.3). All of the hotels and resorts in this

Table 5.3. Breakdown of staff in interviewed hotels by origin (source: authors' own compilation).

	Local	Expatriate
Employees		
Managers (%)	63.1	36.9
Staff (%)	98.1	1.9
Total (%)	94.7	5.3
Average monthly salary		
Managers (US$)	396.56	799.46
Staff (US$)	165.24	412.54
	Male	Female
Gender		
Local managers (%)	44.3	55.7
Expatriate managers (%)	78.0	22.0
Total managers (%)	56.8	43.2
Local staff (%)	46.2	53.8
Expatriate staff (%)	89.5	10.5
Total staff (%)	47.0	53.0

study reported employing the same number of staff in the high season as the low season. In line with the size of the hotels in Bhutan, the average number of managers per hotel is 6.5 (median = 5), ranging from 1 to 14 in the hotels investigated in this research. The average number of staff per hotel is 55.3 (median = 47), ranging from 1 to 173. Including staff and management together, the overall ratio of employees to rooms is 1.87:1. Often, hotels' management teams tend to be predominantly expatriates while staff are sourced locally. In the case of Bhutan, however, 63.1% of managers are local and almost all of the staff (98.1%) are local. The exceptions tend to be several specialized positions such as chefs or swimming pool engineers. This means that, in total, 94.7% of all hotel employees (managers and employees) are Bhutanese. While the number of Bhutanese employed in management positions is in the majority, there is a noticeable difference between the average monthly salaries of local and expatriate employees, both management and staff. While local managers earn US$396.56 per month on average, expatriate managers earn US$799.46 per month on average. A similar pattern emerges for general employees (although there are far fewer expatriate general staff). Local staff earn US$165.24 per month on average, while expatriate general staff earn US$412.54 per month. Staff are also typically paid part of the service charge imposed on tourists. This percentage varies from hotel to hotel, from 0% to up to 80% of a staff member's salary. On average, staff receive an additional 21.9% of their salary (US$36.11 per month) as their share of the service charge.

Bhutan's hotels and resorts generally provide attractive additional employment benefits to their managers and general staff. Both managers and general staff are provided with superannuation entitlements and meal allowances and are paid performance incentive bonuses. Some of the more upscale resorts offer housing allowances and transportation to commute to work.

Given that employees are permanently retained whether it is high season or low season, there are very few casual employees. Only four of the 19 hotels reported having casual employees. These casual employees tended to be local students undertaking internships at these hotels. In this instance, Bhutanese hotels are contributing to the further training and education of the future workforce in the Bhutanese hospitality industry. Swarbrooke (1999, p. 231) notes that the tourism and hospitality industry has traditionally discriminated on the basis of gender. He highlights that, in general, women have rarely been able to break through the 'glass ceiling' to higher management positions. In Bhutan, though, women fill 43.2% of management positions and 53.0% of general staff positions. Among local managers, females comprise the majority of positions but among the expatriate managers, men comprise 78.0% of the positions.

Research participants were asked what percentage of their staff came from relatively poor backgrounds. There is a vast range among the reported numbers, with two establishments reporting that 100% of staff were from poor backgrounds while another hotel reported that no staff were from poor backgrounds as they only employed experienced staff. The average proportion of staff from relatively poor backgrounds is 56.7%. Given the relatively attractive working conditions, it is not surprising that staff turnover is relatively low (for the hospitality sector), at an average of 9.7%, with a minimum of 0% and a maximum of 28%.

5.3.3 Handicrafts/souvenirs

Souvenirs are an important category of products in the tourism context, both as elements of the tourism experience and as a sector of the tourism economy (Jansen-Verbeke, 1998; Swanson and Horridge, 2006). In other developing countries (e.g. The Gambia and Vietnam; Mitchell and Ashley, 2010), the sale of handicrafts is one way that, through the value chain, local artisans can generate income and secure seasonal livelihood opportunities for whole families (Mitchell and Ashley, 2010). The production of handicrafts can support traditional crafts (Markwick, 2001), empower women in local societies (Kruger and Verster, 2001) and engage producers in influencing policy (Makhado and Kepe, 2006). While in other contexts, hotels and resorts provide one type of outlet for these products, in Bhutan hotels and resorts rarely sell handicrafts or souvenirs on their premises. For those hotels in this study that do sell such goods, 100% of the goods are locally produced. Only six of the 19 hotels offer handicrafts or souvenirs for sale. In and around Thimphu and Paro, speciality shops exist to provide tourists with souvenirs. The business model in these stores is that they display the art and handicrafts, and the artists are paid when the tourists actually purchase the goods. In other words, they are paid on consignment. All in all, the sale of handicrafts and souvenirs is an underdeveloped opportunity in the hotels addressed in this study, bypassing possibilities for augmenting the tourist experience of these accommodation establishments and creating local economic linkages.

5.3.4 Transportation

Another avenue that can contribute to poverty alleviation and disperse the benefits of tourism expenditure is local transportation. Transport is a major factor in tourism development and operations (Israeli and Mansfeld, 2003; Page, 2016) that can have a profound impact on the tourist experience (Lumsdon and Page, 2004). The development of transportation has played an important role in enabling the development of tourism and the economies that surround it (Culpan, 1987; Sorupia, 2005). Relying as it does on considerable expenses and the complex coordination of multiple systems, transportation can be a constraining factor in contexts that are still developing. In many developing destinations like Bhutan, the high operating and maintenance costs of vehicles, the lack of domestic vehicle production, high-cost imported fuel, and the wear and tear exacted by rough roads all add to the difficulties of maintaining a transport service (Heraty, 1989). Again, typically in developing countries, such as Tanzania and Ethiopia (Mitchell and Ashley, 2009), acquisition of transport for tourism activities can occur through the value chain, where hotels can work with local taxis and minivan operators to provide transportation services for tourists. However, in Bhutan, while the transportation operators are 100% local, the controlled nature of how tourists visit the mountainous kingdom means that the decision as to which provider to use for transportation is made by the tour operator with which the tourist books the package holiday. This means that hotels in Bhutan are not in a position to forge linkages and synergies with transport providers, as accommodation and transport are two separate modules in the packages put together by tour operators. Additionally, while improvements in transport modes can increase accessibility to a wider range of sites (Lumsdon and Page, 2004), this motivation is also constrained by the Bhutanese government's strict limitation on the locations that tourists are allowed to visit.

5.4 Discussion

Through analysis of interviews with managers of 19 hotels of different classes in Bhutan, this chapter has investigated the roles that these establishments play in the sustainability of tourism in this developing mountainous nation. This research has confirmed that in many ways the characteristics of the Bhutanese accommodation sector correspond to familiar trends in developing countries around the world. For example, pay for expatriates and locals in equivalent jobs in the hotels under study was found to be greatly disparate, and food and beverages is a major factor in economic leakages because of the need to import food to meet the standards, consistency and volume demanded by the tourism sector.

However, there were also some indications of factors in which these hotels defy accustomed characteristics. Hotels in Bhutan were found to differ from the general trend in maintaining their full cohort of employees throughout the year, regardless of the extreme fluctuations of tourist demand between high and low seasons. This contributes significantly to the social sustainability of this sector, as it minimizes the uncertainty, at the household level, that comes with seasonal work. Also, the majority of members of the management teams in the surveyed hotels are Bhutanese rather than expatriates, diverging from the trend present in many developing countries. This factor also contributes to the economic sustainability of the industry from the point of view of the local families and households. It also supports capacity building in terms of building up a knowledgeable and experienced leadership level among people who are likely to remain in Bhutan and apply their knowledge and experience to the domestic hospitality sector, as opposed to expatriates, who would be more likely to remain for a limited time before moving back to their home countries or on to employment at other destinations. However, as mentioned above, there is still an extreme disparity in pay between locals and expatriates in management positions, perpetuating the implicit neocolonialist dual-class system apparent in many developing countries.

In terms of social sustainability, as in many places, the majority of employees in the hotels studied were from poor backgrounds, and the lack of seasonality of employment implies that this is an option for stable incomes for such disadvantaged people. There is more gender equality in management roles among Bhutanese management employees compared with expats in this same group of surveyed hotels, with women accounting for the majority of leadership positions held by Bhutanese, as opposed to only 22% women among expat managers in the sample. The fact that the minority of managers in hotels worldwide are female indicates that the accommodation sector in Bhutan is more of a force for empowerment of women than is typical in other countries – another factor in social sustainability.

Government programmes to educate farmers in producing food suitable for the tourism hospitality sector and training chefs in the use of local ingredients in preparing food suitable for the tourism trade are positive signs that can contribute to alleviating these weaknesses, but such programmes can do little to address the limiting factors of the quantity, price and consistency of domestically produced food, which are all constraining factors on hotels' ability to patronize local farmers. In developed countries, large quantities, low unit prices and high consistency are all attained through the industrialization of food production, which takes control out of the hands of small-scale, local, family-run producers, in addition to being at odds with the Bhutanese government's cautious approach to modernization. Consequently, this ethos can be seen as setting limits to the viable amount of domestic food products used in the hospitality sector.

The long transport routes necessary to bring this food into the country detract from the sustainability of this aspect of the hospitality sector. This is a notable weak link in the government's efforts to minimize the impacts of tourism on the natural environment of the country. The habit of importing food for the tourism industry is also detrimental to economic sustainability, as it is associated with leakages in the tourism economy, displacing economic benefits that could otherwise accrue to domestic producers. Additionally, it represents a significant missed opportunity in terms of social sustainability, as it eschews the opportunity to support local farming livelihoods. Bhutan's unusual relationship to meat is a unique cultural factor that transcends the tourism industry. Demand for meat is driven largely by the domestic market and the meat dishes prepared for tourists are part of the national cuisine and not devised just for the tourist market.

The low number of hotels selling local handicrafts represents a missed opportunity to create linkages and synergies between the handicrafts and hotel sectors. This would seem to have a simple solution – to integrate more shops selling such wares into hotels – but more research is needed to ascertain the reasons why this practice is not common, as well as the institutional barriers that would need to be overcome to enable this. In particular, prior studies have revealed that favourable social and economic impacts for the people and communities that produce souvenirs and handicrafts can best be attained when these producers are engaged as partners in co-developing models for collaboration, rather than as mere providers of goods (Mitchell and Ashley, 2009).

A similar point could be made regarding the lack of linkages between the hotel and transport sectors, although, again, further study and co-development of strategies would be required to define the correct form that such linkages would take in the Bhutanese context. The high degree of control over tourists' itineraries, requiring visitors to travel predetermined routes, patronize preordained establishments and visit only the sanctioned sites, always under the watchful eye of a tour guide, is another manifestation of the Bhutanese government's programme of protectionism that certainly achieves its intended goal of controlling the impacts of tourism, but at the cost of minimizing the spread of potential benefits of tourism.

The class of tourists who visit Bhutan is more affluent and educated than those who make up the mass tourism market that is at the root of the worst impacts of tourism on other destinations. The Bhutanese government could take courage from this fact as an impetus to revisit the ways in which controls are exercised over this sector, and evaluate ways in which it might experiment with allowing greater interpenetration of tourism with other parts of the Bhutanese economy, society and culture.

Conclusions

The above discussion has probed the various ways in which tourism planning and policies in the hotel sector in Bhutan promote, reinforce and constrain sustainability, in its multiple facets. The various factors that contribute to Bhutan's national mantra of Gross National Happiness, informed by the country's deeply ingrained Buddhist values, align well with many elements of the broad concept of sustainability. Preservation of the natural environment and local cultural practices and values, contribution to the local economy and livelihoods, support of opportunities across genders and economic classes, and other factors explored in this chapter have relevance in both of these value systems.

Hotels are the epitome of location-based assets in the tourism ecosystem. They are nodes at which diverse factors and systems of the tourism sector converge – pivotal environments in the tourism experience, innately embedded in the economic, social, cultural and political specificities of their location. As such, they are sites that bear intensive investigation in terms of their potential to positively influence the sustainability of the social values, economies and livelihoods of their localities, while enriching tourists' engagement with the destination. In sum, the findings of this study point to the benefits to be gained, in terms of sustainability and Gross National Happiness, by exploiting hotels' unique position in the tourism landscape. This would require a revisiting of the centrally controlled, tour-operator-led approach to tourism in Bhutan, to divest some degree of decision-making and organization to the local level, enabling localities or even individual accommodation establishments to forge partnerships with local providers of food and beverages, workforce, souvenirs and transport. At the national level, this would need to be supplemented with policies and programmes to help local providers to increase their capacity and capabilities.

The country's cautious protectionist approach to tourism development underlies many of the positive contributions to sustainability identified in this chapter but is also at the foundation of some of the limitations that constrain performance in terms of other measures of sustainability. While these controls certainly shield Bhutanese culture and society from the negative impacts of the worst manifestations of global tourism, such policies also get in the way of the spread of the benefits of tourism across Bhutanese society and impede potential linkages between tourism and hospitality and other sectors. The raising of capacity to forge and maintain such linkages would in many cases entail the modernization and rationalization of elements of the Bhutanese economy to meet the demands of the tourism industry, meaning that the consideration of a path forward in the development of tourism in Bhutan will entail a principled revisiting of the relationship between modernity, tradition, quality of life and sustainability, and the potential policy implications thereof.

There are several limitations to this research that provide areas for future research. Despite the 19 surveyed hotels being relatively representative of the accommodation sector in Bhutan, in absolute terms, future research could undertake a more comprehensive audit of hotels covering accommodation places outside Paro and Thimphu. This research examines the supply side of tourism in Bhutan. Future research could also examine the attitudes, preferences and choices surrounding the issues of sustainability of tourists to Bhutan. The high value of tours to Bhutan and the type of tourism 'product' offered in Bhutan would suggest that tourists may be more 'sustainable'. It would be of interest to understand these tourists better. Lastly, given the importance of tour guides in making sustainable choices, future research could explore the thinking and decision making of these important tourism stakeholders around the issue of sustainability.

References

Bates, W. (2009) Gross national happiness. *Asian-Pacific Economic Literature* 23(2), 1–16.

Bélisle, F.J. (1983) Tourism and food production in the Caribbean. *Annals of Tourism Research* 10(4), 497–513.

Bramwell, B. and Lane, B. (1993) Sustainable tourism: an evolving global approach. *Journal of Sustainable Tourism* 1(1), 1–5.

Brunet, S., Bauer, J., De Lacy, T. and Tshering, K. (2001) Tourism development in Bhutan: tensions between tradition and modernity. *Journal of Sustainable Tourism* 9(3), 243–263.

Culpan, R. (1987) International tourism model for developing countries. *Annals of Tourism Research* 14(4), 541–552.

Dema, K. (2015) No slaughterhouse, govt. clarifies. Available at: https://kuenselonline.com/no-slaughterhouse-govt-clarifies/ (accessed 4 February 2020).

Dorji, T. (2001) Sustainability of tourism in Bhutan. *The Journal of Bhutan Studies* 3(1), 84–104.

Goeldner, C.R. and Ritchie, J.R.B. (2012) *Tourism: Principles, Practices, Philosophies*, 12th edn. Wiley, Hoboken, New Jersey.

Gurung, D.B. and Seeland, K. (2008) Ecotourism in Bhutan: extending its benefits to rural communities. *Annals of Tourism Research* 35(2), 489–508.

Heraty, M.J. (1989) Tourism transport – implications for developing countries. *Tourism Management* 10(4), 288–292.

Hummel, J. and van der Duim, R. (2012) Tourism and development at work: 15 years of tourism and poverty reduction within the SNV Netherlands Development Organisation. *Journal of Sustainable Tourism* 20(3), 319–338.

Israeli, Y. and Mansfeld, Y. (2003) Transportation accessibility to and within tourist attractions in the Old City of Jerusalem. *Tourism Geographies* 5(4), 461–481.

Jansen-Verbeke, M. (1998) The synergism between shopping and tourism. In: Theobald, W. (ed.) *Global Tourism*, 2nd edn. Butterworth-Heinemann, Oxford, pp. 428–446.

Kruger, S. and Verster, R. (2001) An appraisal of the Vulamehlo handcraft project. *Development Southern Africa* 18(2), 239–252.

Liu, Z. (2003) Sustainable tourism development: a critique. *Journal of Sustainable Tourism* 11(6), 459–475.

Lumsdon, L. and Page, S.J. (2004) *Tourism and Transport: Issues and Agenda for the New Millennium*. Elsevier, Oxford.

Makhado, Z. and Kepe, T. (2006) Crafting a livelihood: local-level trade in mats and baskets in Pondoland, South Africa. *Development Southern Africa* 23(4), 497–509.

Markwick, M.C. (2001) Tourism and the development of handicraft production in the Maltese islands. *Tourism Geographies* 3(1), 29–51.

Meyer, D. (2007) Pro-poor tourism: from leakages to linkages. A conceptual framework for creating linkages between the accommodation sector and 'poor' neighbouring communities. *Current Issues in Tourism* 10(6), 558–583.

Mitchell, J. (2013) Value chain approaches to assessing the impact of tourism on low-income households in developing countries. *Journal of Sustainable Tourism* 20(3), 457–475.

Mitchell, J. and Ashley, C. (2009) Value chain analysis and poverty reduction at scale. Overseas Development Institute Briefing Paper 49. Available at: http://www.odi.org.uk/resources/docs/3528.pdf (accessed 7 July 2012).

Mitchell, J. and Ashley, C. (2010) *Tourism and Poverty Reduction: Pathways to Prosperity*. Earthscan, London.

Nyaupane, G.P. and Timothy, D.J. (2010) Power, regionalism and tourism policy in Bhutan. *Annals of Tourism Research* 37(4), 969–988.

Page, S.J. (2016) *Transport and Tourism*. Pearson Education Limited, London.

Pillay, M. and Rogerson, C.M. (2013) Agriculture–tourism linkages and pro-poor impacts: the accommodation sector of urban coastal KwaZulu-Natal, South Africa. *Applied Geography* 36, 49–58.

Pratt, S., Mackenzie, M. and Lockwood Sutton, J. (2017) Food miles and food choices: the case of an upscale urban hotel in Hong Kong. *Journal of Sustainable Tourism* 25(6), 779–795.

Pratt, S., Suntikul, W. and Dorji, U. (2018) Economic sustainability? Examining the linkages and leakages between agriculture and hotels in Bhutan. *International Journal of Tourism Research* 20(5), 626–636.

Rinzin, C., Vermeulen, W.J.V. and Glasbergen, P. (2007) Ecotourism as a mechanism for sustainable development: the case of Bhutan. *Environmental Sciences* 4(2), 109–125.

Ritchie, M. (2008) Tourism in the Kingdom of Bhutan: a unique approach. In: Cochrane, J. (ed.) *Asian Tourism: Growth and Change*. Elsevier, Oxford, pp. 273–283.

Royal Government of Bhutan (2005) *Sustainable Tourism Development Strategy*. Royal Government of Bhutan, Thimphu.

Sharma, M. and Zam, K. (2019) Historical development of Vajrayana Buddhism in Bhutan. *Our Heritage* 67(2), 875–883.

Shrotryia, V.K. (2006) Happiness and development – public policy initiatives in the Kingdom of Bhutan. In: Ng, Y.-K. and Ho, L.S. (eds) *Happiness and Public Policy: Theory, Case Studies and Implications*. Palgrave Macmillan, London, pp. 193–208.

Smith, V.L. (2012) *Hosts and Guests: The Anthropology of Tourism*, 2nd edn. University of Pennsylvania Press, Philadelphia, Pennsylvania.

Sorupia, E. (2005) Rethinking the role of transportation in tourism. *Proceedings of the Eastern Asia Society for Transportation Studies* 5, 1767–1777.

Suntikul, W. (2018) Cultural sustainability and fluidity in Bhutan's traditional festivals. *Journal of Sustainable Tourism* 26(12), 2102–2116.

Suntikul, W. and Dorji, U. (2016) Local perspectives on the impact of tourism on religious festivals in Bhutan. *Asia Pacific Journal of Tourism Research* 21(7), 741–762.

Swanson, K.K. and Horridge, P.E. (2006) Travel motivations as souvenir purchase indicators. *Tourism Management* 27(4), 671–683.

Swarbrooke, J. (1999) *Sustainable Tourism Development*. CAB International, Wallingford, UK.

Tourism Council of Bhutan (2019) Bhutan Tourism Monitor: Annual Report 2018. Available at: https://www.tourism.gov.bt/uploads/attachment_files/tcb_xx8r_BTM%202018%20_final.pdf (accessed 2 February 2020).

Ura, K., Alkire, S., Zangmo, T. and Wangdi, K. (2012) A Short Guide to Gross National Happiness Index. Available at: https://opendocs.ids.ac.uk/opendocs/bitstream/handle/20.500.12413/11807/Short-GNH-Index.pdf (accessed 11 March 2020).

6 Tourism Policy Planning in Bangladesh: Perspective and Propositions

Azizul Hassan*

Tourism Consultants Network, The Tourism Society, UK

Introduction

Effective tourism policy planning is a facilitator for promoting economic growth in many developing countries in Asia and the Pacific (Asian Development Bank, 2016). The success of tourism policy planning depends on many relevant factors including the participation of both the public and private sectors (Andriotis *et al.*, 2018). At the national level, policy makers are expected to highlight tourism priorities, focusing on macrolevel impacts and taking account of the interests of a wide range of stakeholders. In general, policy planning is the process of decision formation having a clear vision of implementation followed by monitoring and systemic feedback mechanisms (Airey and Ruhanen, 2014). In this sense, the government has specific roles to perform to ensure the maximum output from any sector, including tourism. Tourism policy planning includes regulations, rules, guidelines, directives and development/promotion objectives and strategies that provide a framework within which the collective and individual decisions directly affecting tourism development and the daily activities within a destination are taken (Ritchie and Crouch, 2003). Bangladesh, as the case of research presented in this chapter, possesses the natural resources, human capital and economic potential to develop a tourism industry which can attract international tourists (Hassan and Burns, 2014; Hassan *et al.*, 2020). This research critically explains tourism policy planning in Bangladesh (i.e. tourism policy planning, instruments, organizational involvement and implementation). Primary data allowed access to the relevant policy planners', stakeholders' and beneficiaries' opinions. Thus, the research has uniquely incorporated these opinions with the conceptual discussions. Based on the findings, this chapter generates some specific suggestions for tourism policy planning in Bangladesh.

*E-mail: azizulhassan00@gmail.com

DOI: 10.1079/9781789246698.0006

This chapter is organized with consistency and coherence. A theoretical overview of tourism policy planning comes at the beginning, followed by an explanation of policy instruments, tourism policy planning and organizational involvement, policy implementation and Bangladesh tourism policy. The methodology section is immediately followed by findings and analysis. The conclusion and recommendations section summarizes the research findings against the aim and objectives of the research and offers some suggestions.

6.1 Tourism Policy Planning

6.1.1 A theoretical overview

The theoretical analysis of tourism policy planning in this section of the research covers tourism policy planning stages, power relations, role identification of the relevant parties, relevant policy considerations, sustainability and other concerns, collaboration and partnership, and policy outcome consideration (Fig. 6.1).

Governance

Jamal and Camargo (2018) find that justice is a leading principle for good governance and policy in tourism; as well as the importance of cultural values,

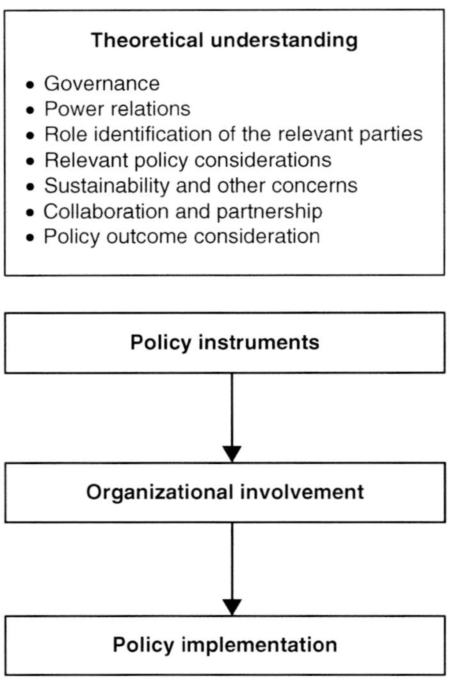

Fig. 6.1. Theoretical framework (source: author's own elaboration).

addressing the historical context is also relevant. In supporting governance, Seonyoung and Bramwell (2019) emphasize varied boundary relations in society, including political, geographical, social and discursive relations. Edgell and Swanson (2018) describe the logical sequences for tourism policy planning: clarity of goals and objectives (i.e. deciding what a country or region wants from tourism and what it is willing to invest and sacrifice), identification of tourism issues, qualitative and quantitative research to create baselines for measuring progress, understanding the range of impact potentials (including on society and the environment), who and what shape and influence tourism issues, and finally policy formation, decision and implementation, with linked evaluation, monitoring and feedback systems.

Power relations

The development of tourism policy is closely related to power relations within a state (Bowen *et al.*, 2016). Centralized tourism policy followed by hierarchy of organizations and agencies should also be given emphasis because the implementation plan requires statutory involvement of local government agencies, like in Thailand (Krutwaysho and Bramwell, 2010). Tourism policy planning in many cases is manifested in political party agendas as well as in the government's political views (Chambers and Airey, 2001). Political debate, like ethnic rivalry in Indonesia, has been ignored in terms of tourism policy formulation (Chheang, 2009). Also, regional coordination for tourism endorsements can become a challenge. An example was the East African Community (EAC), which was unable to reach the objectives due to conflict of interests (Kagwanja, 2007). Power relations have a certain influence in national tourism policy planning.

Role identification

Effective policy planning will identify the respective roles of government, private and third sectors in defining and developing the tourism system and its supporting structures (Alavi and Yasin, 2000). The roles and responsibilities of the national tourism organization (NTO) in the national tourism policy need to be properly identified. As a relevant example, Bangladesh Parjatan Corporation (BPC), as an autonomous organization under the Ministry of Civil Aviation and Tourism, was formed in November 1972 and started its activities in 1973. Many countries have corporatized the NTO, which has forced this organization to become profit-oriented rather than service-oriented. Role identification helps prevent unexpected interventions of relevant stakeholders and allows agencies to work more independently (Tuhin and Majumder, 2011).

Relevant policy considerations

Tourism policy planning needs to be integrated with other relevant policy considerations. At the national level, tourism policy planning is attached to, and integrated with, wider policy areas driven by financial aspects such as the level of investment commitment from government related to potential revenue and employment gains. In this sense, 'tourism is intricately connected and cannot be separated from other policy areas' (Stevenson *et al.*, 2008, p. 745). Hall (2000) reinforces the idea that tourism policy can be integrated with other policy areas.

Sustainability and relevant concerns

With specific reference to the stages of strategic tourism planning processes, Gunn (2002) emphasizes the need for establishing terms of reference, taking an environmental approach and consulting stakeholders. Butler (2009) argues that countries which are still at what is often called the 'take off' stage of tourism should focus on effective and sustainable tourism policy implementation to create the ground for service-oriented tourism rather than going for profitability and equitable distribution of benefits (economic, social and regional) from the very beginning. Apart from this, a number of endogenous and exogenous factors can arise that, in the first instance, seem beyond the control of tourism planners. For example, COVID-19, severe acute respiratory syndrome (SARS), swine flu and foot-and-mouth disease have wreaked havoc on international mobility and caused temporary devastation in destinations and among tourism businesses (Telfer and Hashimoto, 2015; Kreiner and Ram, 2020).

Collaboration and partnership

Collaboration and partnership are another aspect of tourism policy planning and implementation. Selin (1999) reminds that collaboration and partnership in tourism can provide the base to understand that stakeholder networking and joint effort highly prioritize the development and sustainability of tourist destination through policy and planning. This latter point is made by Dredge (2006), as networks spanning private and public sectors are becoming more important to shape tourism planning and development. Policy planning relies on capacity betterment of public agencies in terms of research and information sharing through collaboration and partnership.

Policy outcome consideration

According to Telfer and Sharply (2016), tourism helps to increase income, foreign exchange, employment and government revenues, and boosts other subsectors through value-and-supply chain activities. In the societal and cultural context, tourism plays a role in educating people, delivering cross-cultural exchange, and, at a more practical level, offering extensive choices of recreation. Hall (2009) argues that tourism policy is essential to ensure a workable system that improves the general quality of tourism, from both the 'market' and 'product' perspectives. Tourism policy is important among all sectors and to guide national tourism to achieve certain successes. As viewed by Zhang *et al.* (1999), in China, tourism policy is offered enormous importance to create the ground of efficiency in tourism-related administrations, infrastructures, managements, and product or service offers. Policies can help tourism to contribute more to the national gross domestic product.

6.1.2 Policy instruments

In a discussion on policy instruments and actions, Dredge and Jenkins (2007, p. 181) note that 'policy instruments are best understood as the tools and approaches used by governments to achieve outcomes and might be broadly classified according to the types of resources they use (e.g. information, money, government action, and law)'. Clearly, the types of policy instruments depend on the nature of policy problems, measurement and the level of government involvement. But whatever the circumstances, such instruments have to be seen as mechanisms of resource deployment. Hood (1986) offered a typology of policy instruments to be adopted by government. He divided them into four groups: information instruments, treasury instruments, authority instruments and organization instruments. Policy instruments can be set in accordance with implementation channels and the level of government involvement. One powerful policy instrument is the authority instrument, which includes legislation, regulation and enforcement. Recent trends of sexual exploitation and trafficking of children and women are just such an area that requires the exercise of governmental authority instruments. Dredge and Jenkins (2007) again identified four factors that influence the choice of policy instruments: the disposition of the state, knowledge of the policy instruments available, current events and knowledge of the consequences. The choice of policy instrument is important because of the need for successful implementation of tourism policy. The choice and use of policy instruments vary across individual country contexts, including the extent to which countries can quickly enact legislation, the extent of political willingness to support tourism's growth and the clarity of policy objectives.

6.1.3 Tourism policy planning and organizational involvement

The potential outcomes of various tourism planning approaches are influenced by society and community development, private and public stakeholders' involvement and individual countries' interest for economic benefits. Thus, tourism planning should be seen as non-static and subject to the monitoring and feedback mechanisms mentioned above. Tourism policy depends largely on multilevel organizational involvement. At the international level of tourism, organizations consider the macro-environmental trends of a specific region or country while formulating or assisting tourism policy planning. The United Nations World Tourism Organization (UNWTO) is the specialized agency and the leading international organization in the field of tourism. It serves as a global forum for tourism policy issues and a practical source of tourism know-how (UNWTO, 2020). The World Travel & Tourism Council (WTTC) is an awareness-raising global forum for travel and tourism business leaders, while a major contribution of the WTTC is the Tourism Satellite Accounting (TSA) research-based country profile, which claims that tourism's economic contribution in Bangladesh has been undercounted and underrated (WTTC, 2020).

6.1.4 Policy implementation

Tourism policy planning is complex, sometimes disputed, but whatever the prevailing political and economic circumstances, it should be directly intertwined with implementation: the process through which policy ideas and plans are translated into practice (Telfer and Sharply, 2016). The implementation of tourism policy (which itself frames demand, supply and organization development) depends on the effectiveness of local, regional and national officials working in partnership with the private sector and other stakeholders, including the host populations and environmental and cultural third-sector/lobbying/protection groups (Dodd and Butler, 2009). Tourism policy implementation can be very difficult in many cases (Lai *et al.*, 2006). In this sense, the public sector is motivated by economic, social and environmental concerns and provides the political will and practical facilitation framework for the tourism sector, which, in turn, is driven by understanding tourist needs and wants. In developing countries, tourism policy implementation is regarded as a challenge because of the wider gap between plan and action (Kilicbeyli, 2011). This is why, in recent times, emphasis has been given to implementing tourism policy more effectively. In this regard, three archetypal approaches of policy implementation as suggested by Hall (2009) can be seen as valid. Out of these approaches, the 'top down' and 'bottom up' approaches are found as beneficial for tourism policy implementation in developing countries (Andriotis, 2018). The 'top down' approach follows the hierarchy sequence from top-level policy planners to local level, while the 'bottom up' approach reflects a low level of coordination between the top and local levels. Implementation aspects should comply with the key questions raised by Hall (2009), where a policy–action relationship is prioritized to chalk out resolution of key challenges.

6.1.5 Bangladesh tourism policy

Before 1992, Bangladesh had no tourism policy but rather a strategic Master Plan for tourism development prepared by the United Nations Development Programme (UNDP) and UNWTO in 1990. The first tourism policy was adopted by the government in 1992. Following the process and the global demand, the government adopted an updated tourism policy in 2009. The National Tourism Policy of Bangladesh was issued, published and in force on 14 December 2009. The responsible ministry and governmental agencies for implementing the policy are the Ministry of Civil Aviation and Tourism and BPC, and Bangladesh Tourism Board (BTB), respectively (Hassan and Burns, 2014). The updated tourism policy of Bangladesh is a comprehensive tourism policy setting out the goals, objectives, policy instruments, implementation aspects and evaluation criteria. The Tourism Policy of Bangladesh 2009 has 31 objectives and goals with brief policy actions. Before moving on to details of the policy, presenting its key points is useful.

The Bengali version of the National Tourism Policy of Bangladesh is presented on the official website of the Ministry of Civil Aviation and Tourism but can hardly be read by any foreigner due to language issues (Ministry of Civil Aviation and Tourism, n.d.). Also, this important document is not easily accessible to download, suffering from technical issues on this website (Ministry of Civil Aviation and Tourism, 2020). In addition, the National Tourism Policy is also available on the official website of the BPC. The tourism policy as uploaded on the website has year 2009 mentioned in the heading, but it says the national policy was formulated in 1992. Thus, the contents are mixed up with one another and make it unclear which year's tourism policy it is (BPC, n.d.). When developing countries are adopting multilingual websites in the interests of meeting the needs of diverse domestic and international audiences, this ambiguity creates a ground for conflict between international and local experts and provides complexity in the analysis of tourism policy (Hassan and Kokkranikal, 2018).

At the time of drafting the policy, the concerned authorities should have clear knowledge about global trends and use the most recent socio-political and socio-economic data available. It is not at all clear if this was the case in Bangladesh, due to the lack of published material on the Bangladesh tourism policy planning process. Kennell and Chaperon (2013) have noted that the same lack of information causes difficulties in the analysis of the UK's most recent tourism policy. Policy planning that takes place without sufficient analysis of data and information can create policy complexities. An example is that, until 1997, there were only two World Heritage Sites in Bangladesh (the historic Mosque City of Bagerhat and the ruins of the Buddhist Vihara at Paharpur) (World Heritage Sites, 2020). For both these sites, there was a significant lack of information and data for undertaking further development and conservation initiatives. Unexpectedly, the 'Summaries of Periodic Reports Submitted by States Parties – 2003' could not provide any statistical data on visitors to both World Heritage Sites. However, this lack was clearly mentioned in the 'State of Conservation of the World Heritage Properties in the Asia-Pacific Region', resulting in complications and even financial cuts in the development programmes (United Nations Educational, Scientific and Cultural Organization (UNESCO), 2010). Evidently, the lack of data on either site led to damaging policy decisions, including a lack of future funding. Both BPC and BTB, as the NTO, appear unable to provide updated statistical data on their official websites or in their various reports (BTB, 2020). Identifying the priority sectors is important to ensure the maximum possible outcomes. Evidently, 'Bangladesh, for example, as a riverine country is considering developing river boat tourism' (Inskeep, 1991, p. 165). Or, according to Bangladesh Caterers Association (BCA, 2019), in the UK nearly 12,000 British Bangladeshi-owned restaurants are in operation employing 100,000 people and with an annual turnover of £4.2 billion.

6.2 Methodology

Narrowing the 'gap' between policy makers and people has never been straightforward as long as the distancing effects are produced by a combination of knowledge, geography and politics. Therefore, this research is partially based on 'methodological populism' as suggested by Lewis (2018), which combines participatory and ethnographic approaches that challenge the present managerial cultures of what can be considered as acceptable evidence for policy. In line with 'methodological populism', data and information were collected from in-depth and face-to-face interviews and open-ended discussions. Twenty face-to-face interviews were conducted with three sets of interviewees with no access issues. The first set of respondents comprised professionals working in private tourism enterprises (i.e. travel agents, tour operators, hotels and restaurants). These respondents somehow play a role in the national tourism policy planning process with suggestions from their standpoint. These respondents as stakeholders tend to be affected by tourism policy. The second respondent set involved officials in the public sector involved in tourism policy planning. The third group comprised general tourists. These respondents were not involved with any formal policy planning activities but represent the subjects to be affected directly by national tourism policy. Only domestic tourists were selected as general respondents.

These interviews and open-ended discussions reflected the policy framework of tourism planning covering the private, public and individual viewpoints and ideologies. The basic query to all of the respondents was their opinion about the existing tourism policy and policy suggestions. These interviews and discussions very specifically covered theoretical aspects of tourism policy planning as outlined in the theoretical framework (Fig. 6.1), exploring the respondents' views about tourism policy planning, their expectations and their suggestions to overcome issues and challenges. The sample respondents were selected purposively. These interviews were conducted in May 2019 as follow-ups to initial interviews with similar respondents (i.e. although not the same people) in 2016. The rationale for such selection was to reveal ideas from the private sector, government and beneficiaries such as general tourists. The face-to-face interviews were audio-recorded. The interviews were undertaken in Bengali and then translated into English and self-transcribed. This allowed the researcher to understand the contents of the interviews and discussions more intensively. For interpretation and content analysis, the first set of respondents was labelled as A1 to A7, the second set as B1 to B6 and the third set as C1 to C7 (Table 6.1). The first set of respondents were respondent category 'A' (i.e. professionals of private tourism enterprises), the second set of respondents category 'B' (i.e. public-sector tourism policy planners) and the third set of respondents category 'C' (i.e. general tourists).

6.3 Findings and Analysis

Findings reveal issues relevant to the tourism policy planning in Bangladesh.

6.3.1 Effectiveness of the present tourism policy in Bangladesh

Face-to-face interviews with the respondents carefully covered areas mentioned in the literature review section. Respondent A2, as the representative of the Association of Travel Agents of Bangladesh (ATAB), cited collaboration and partnership aspects of the National Tourism Policy of Bangladesh as invalid. On the contrary, A1, A3 and A4 viewed the policy as excellent. A5 expressed that 'the national tourism policy that Bangladesh currently has at least offered some

Table 6.1. Respondents' profile (source: author's own compilation).

Type of respondents	Interview code	Representation
One representative from the Bangladesh Tourism Association (BTA)	A1	Private tourism enterprise professionals
One representative from the Association of Travel Agents of Bangladesh (ATAB)	A2	
Two representatives from the Tour Operators Association of Bangladesh (TOAB)	A3, A4	
One representative from the Bangladesh International Hotel Association (BIHA)	A5	
Two representatives from the Department of Tourism and Hospitality Management, University of Dhaka	A6, A7	
One official from the Prime Minister's Office, Bangladesh	B1	Public-sector tourism policy planners
One official from the Ministry of Planning, Bangladesh	B2	
One official from the Ministry of Civil Aviation and Tourism, Bangladesh	B3	
One official from the Bangladesh Parjatan Corporation (BPC)	B4	
Two officials from the Bangladesh Tourism Board (BTB)	B5, B6	
Three male domestic tourists	C1, C2, C3	General tourists
Three female domestic tourists	C4, C5, C6	
One under-18 domestic tourist	C7	

scopes to collaborate and making partnerships with expatriate investors that results in building so many good-quality hotels and restaurants'. Supporting the statement of A5, respondent A3 stated that 'the tourism policy of Bangladesh has international standard. I frequently work with tour operators of other countries to attract tourists to this country'. Respondents A6 and A7, representing the Department of Tourism and Hospitality Management of the University of Dhaka, strongly criticized bureaucracy and the government's inattention to diversifying the benefits of tourism. These respondents, at the same time, stressed reducing the power exercise of the ruling political party in both tourism policy planning and implementation.

The next respondent group was the public-sector tourism policy planners, dealing with the national tourism policy planning directly. Respondent B3 clearly stated that 'in principle, we are designated to work for the national tourism policy planning. We have to have competency and sufficient knowledge for this, and we follow all stages of tourism policy planning'. Respondents B4, B5 and B6 were similar in their opinion, while respondent B6 argued the same, that 'we are the NTO of Bangladesh and thus responsible for the national tourism policy planning and promoting the tourism industry of Bangladesh'. Regarding relevant concerns such as sustainability or security, respondent B1 informed that 'I work at the Prime Minister's Office and we have clearly instructed relevant government agencies to find solutions to these issues in the national tourism policy planning process'. Respondent B2 as the representative of the Ministry of Planning supported the statement of B1 and informed that they have to reflect directions from the government to ensure that plans are feasible and achievable.

The respondents were invited to comment on policy outcome consideration of the existing tourism policy of Bangladesh. The respondents showed diversity in their views and opinions. In the collective opinions of respondents, C4, C5, C6 and C7 informed that they could not find any outcome that the existing tourism policy of Bangladesh can offer. C1 argued that 'Bangladesh tourism policy has never been effective and timely. These policies since the independence of Bangladesh in 1971 have completely failed to leave any positive outcome on the tourism industry. Whatever the development is made in the tourism industry is sporadic and unplanned'. Respondent C2 mentioned the bureaucratic issues and political instability that negatively affects tourism. Respondent C3 demanded that 'the national tourism policy of Bangladesh needs to come out of political biasness and bureaucratic difficulties'.

Discussions on this point highlight the mixed responses from the respondents. However, the trend of dissatisfaction of the respondents demonstrates the requirement for a valid national tourism policy in Bangladesh. It is important to note that a few of the private-sector tourism professionals viewed the existing tourism policy as uncongenial and unsupportive of development of the tourism industry in Bangladesh. One of the reasons for this can be that tourism in Bangladesh moved without policy before 1992. Even though the public-sector tourism policy planners were typically assertive in their opinions, it was the private-sector tourism professionals who actually had to pass through challenging times.

6.3.2 Tourism policy instruments in Bangladesh

Respondents A1, A2, A3 and A4 demanded the freedom and wider access of private-sector investors and entrepreneurs as the instrument of tourism policy planning. Respondents A1, A2 and A3 again pointed out that available capital investments and tourism-friendly policy are the two main tourism policy instruments that will be made more available in coming years. Respondent A5 stated that 'we want policy instruments that will help expanding our business'. On the other side, respondent A6 informed that 'private-sector tourism should be given policy supports and access to capital loans so we can facilitate the development of tourism activities'. Respondent A6 also recognized applying marketing components like targeting, researching and benchmarking as tourism policy instruments that the government prioritizes for the tourism industry. Respondent A7 identified a few instruments that need to have a place in the national tourism policy planning: training, education, capacity betterment of public agencies, development of information and communication channels, and strategic integration of both private and public sectors. All of the respondents mentioned that the present tourism policy instruments in Bangladesh are somewhat authoritarian instruments supported by regulation and enforcement.

In reference to tourism policy instruments, respondent B1 cited the importance of tourism policy, legislation and implementation capacities. Respondent B2 stressed on ensuring the availability of funds as an instrument at the same time as tourism policy planning. Respondent B3 mentioned that 'if you follow the tourism policy of 2009, we have clearly stated that the government would welcome private-sector investments as the instrument of tourism policy planning'. Respondents B4, B5 and B6 mentioned that the NTO should work with the government to support the plan to develop tourism in remote islands like Saint Martins, Nijhum Dwip, Maheskhali and many more. This can act as the instrument to develop tourism in these islands.

Respondent C1 added that 'both the private and public sector needs to cooperate and coordinate to allure more investments to develop tourism infrastructures in Bangladesh with effective policy instruments'. Respondents C4, C5, C6 and C7 supported this opinion. C2 focused on tourism products and services diversification as the instrument. This respondent identified better tourism services and products, organizing fairs and festivals, infrastructural developments and bringing equality in resource distributions as the useful instruments that the national tourism policy should include. Respondent C3 viewed tighter safety and security of tourists as main concerns, and for the purpose suggested democratic practices, legislation and their applications as policy instruments.

6.3.3 Tourism policy planning implementation

On the point of implementing the existing tourism policy in Bangladesh, all of the respondents emphasized proper implementation of any policy. However, their individual views were diverse. Respondents A1, A2, A3 and A5 clearly accepted that the implementation of tourism policy is far more important than planning. Respondent A6 mentioned that 'the implementation of the national tourism policy mostly relies on capacities of the private and public sector. In case both sectors can work together, the policy will be implemented successfully'. Respondent A4 also supported this opinion by stating that 'tourism policy planning is rather an official job, but its implementation is difficult. The government can employ private sector in this implementation process'. Respondent A7 from the expert viewpoint stated that the tourism policy implementation in Bangladesh should involve the opinion and participation of locals. In reality, this is not happening – rather the tourism policy is somewhat imposed from the government, in the 'top down' approach.

Respondents from the public-sector policy planners were convinced that the government has the capacity to implement the national tourism policy. Respondents B1, B2 and B4 strongly mentioned that Bangladesh has a valid tourism policy and that this policy is already on the way of implementation. Respondent B3 stated that 'if you think carefully, the national tourism policy of Bangladesh is an essential document. We are now ensuring its complete and successful implementation in the years to come'. Respondents B5 and B6 as representatives of the NTO also observed that Bangladesh currently has a tourism policy that is completely implementable.

Respondents C4, C5, C6 and C7 had no detailed idea about the procedures of implementing the national tourism policy. Respondent C1 opined that 'I am unsure whether Bangladesh has any capacity to develop the tourism industry with the existing policy'. Respondent C2 was responsive and stated that 'the national tourism policy of Bangladesh is very crucial and unquestionably important to implement', arguing that 'we will always demand for the tourism policy in Bangladesh that is compassionate for our initiatives towards the well-being of tourism'. This view was supported by respondent C3, who argued that 'Bangladesh still stays far behind from other countries regarding the implementation of its national tourism policy. However, the scenario can change with the intervention of both public- and private-sector tourism organizations'.

6.3.4 Tourism policy planning organizational involvement

The first respondent group, of professionals of private tourism enterprises, obviously favoured the involvement of private organizations in the national tourism policy planning. Respondents A1, representing the Bangladesh Tourism Association (BTA); A3, representing the ATAB; and A4, representing the Tour Operators Association of Bangladesh (TOAB), argued that tourism planning needs to get suggestions from private organizations including stakeholders, interest groups, NGOs and relevant beneficiaries. Respondent A7 was straightforward and mentioned that 'tourism organizations in the private sector are required to share ideas, experience and knowledge with the government. These will certainly have positive effects in the national tourism policy'. However, respondent A2 criticized the government mindset and mentioned that 'in almost all cases, we remain unwelcomed by the relevant government officials for sharing our knowledge and experiences'. Respondent A5 mentioned that there are already private-sector tourism enterprises willing to support tourism and hospitality policy planning. However, the scope of their participation is limited. Respondent A6 very clearly stated that the national tourism policy planning in Bangladesh is manipulated by the governmental agencies, with very little or almost no access to participation. Some donor and international agencies are involved subject to project funding. In general, the participation of NGOs, interest groups and other agencies is mostly ignored.

The second group of respondents expressed their opinions on this particular point. Respondents B2, B4, B5 and B6 highlighted the importance of the UNWTO's guidelines and involvement in the national tourism policy planning process. Respondent B3 argued that 'With my considerable years of experiences in tourism, I strongly stress the active involvement of UNESCO and UNWTO in the national tourism policy planning. This will really help'. Respondent B1 opined that 'in most cases and under many circumstances, we do not consider pressure groups' opinions and interests, as the common interest of the country is important'.

On this specific point of tourism policy planning organizational involvement, few notable remarks were generated from the third respondent group. This was possibly the outcome of their representation as general tourists.

Respondent C1 demanded the involvement of NGOs, interest groups and pressure groups in the national tourism policy planning to ensure that the policy becomes representative and covers the interests of all. Three respondents answered this question. Respondents C4, C5 and C6 insisted on the involvement of local residents' knowledge. Opinions of respondents C2 and C7 also fully supported this view. Respondent C3 mentioned that 'the opinions of local people must have to be reflected in the national tourism policy. Such involvement would narrow the gap between themselves and the policy planners'.

6.3.5 Bangladesh tourism policy discussions

Priorities in tourism policy in Bangladesh

This point is rather important in the perspective of this research. The national tourism policy needs to identify priority areas and plan policies that are beneficial for their development. The first respondent group was clear about selecting the priority areas. Respondents A1, A2 and A4 blamed the public-sector policy planners for failing to identify priority areas and plan accordingly. A5 was more focused on plans to develop the Bangladeshi cuisine and stated that 'the authentic cuisine of Bangladesh was relatively unknown even few years ago, but the situation has now changed. Bangladeshi cuisine is now getting popularity in line with the Indian cuisine in Europe and rest of the world'. Respondents A6 and A7 highlighted that the existing national tourism policy has partly identified priority areas but needs a more detailed working plan. Respondent A3, from TOAB, mentioned that niche tourism sectors need to be given proper attention for the development of a beneficial tourism policy planning. Respondent A6 mentioned applying a tourist zoning system to minimize possible harmful effects resulting from mass tourism or over-tourism.

Respondent B1 was straightforward in assessing the government's success in identifying priority areas. Respondent B2 pointed to security and safety issues as the most priority area. Supporting this claim, respondent B3 mentioned that 'objective 27 of the latest Bangladesh policy stated to ensure overall security of the tourists'. Respondents B5 and B6 were responsive towards the argument of B1. Respondent B1 also commented that 'we are successful in identifying priorities to promote the industry of Bangladesh to a further height'. Respondent B3 was more theoretical: 'we are bound in many cases and we have to ensure sustainability in tourism. We have to follow guidelines of agencies like the United Nations Environmental Programme (UNEP) and many more agencies for the purpose. This should not mean that we discourage private sectors'. Respondent B3 again stated that 'building partnerships with parties involved in tourism activities and focusing on betterment of safety and security also require priority. Tourism product and service offer diversification are challenges of policy planning and implementations. As well as tackling climate change effects is also important due to sustainability of natural tourist destinations'.

Regarding the priority areas, the respondent group of general tourists were clearer. Respondent C1 marked out some priority areas and stated that 'gastronomy tourism is becoming widely popular concept all over the world based on Bangladeshi-owned restaurants' success in Europe and mainly the United Kingdom'. Respondents C2, C3, C4 and C7 jointly identified some priority areas and locations like Chittagong and Sylhet hill tracts, Cox's Bazar and Kuakata sea beaches for adventure tourism, ecotourism, canopy tourism, snorkelling and teenager-friendly tourism product and service offers. Respondent C6 added religious tourism with the Bishwa Ijtema, the world's second-largest Muslim gathering. In addition, this respondent added the Islamic cultural heritage sites with the historic Mosque City of Bagerhat. Respondent C5 cited agricultural tourism, festival tourism, medical tourism, island or river tourism, and indigenous tourism as priority areas that can appear as areas of precedence.

Summarizing the above discussions, primarily, the existing Bangladesh tourism policy has not been able to clearly classify prioritized policy areas, nor have even any directions been made for the purpose. Tourism product diversification in Bangladesh is a possibility (Inskeep, 1991).

Respondent suggestions for tourism policy in Bangladesh

Respondents A1, A4 and A5 suggested creating the ground of a congenial tourism market for private-sector tourism professionals in Bangladesh. Respondent A2 from the ATAB pointed to the active participation of private-sector tourism professionals that helps the entire industry to move forward. Respondent A3 clearly mentioned that 'setting up specialized institutions financed by the private tourism entrepreneurs for creating expert tourism professionals is important'. This respondent termed such proposed institutions as a 'Bangladesh Institute of Tourism and Hospitality (BITH)' to be set up in all district headquarters. Respondent A5 asserted on bringing into play the academic and practical excellences in the current National Hotel and Tourism Training Institute, Dhaka. This respondent informed that private-sector investors are already investing in catering and hospitality training. Respondent A7 was more conversant and asserted to introduce a 'Research, Monitoring and Evaluation Wing' and a 'Media and Communication Wing' at the BPC and BTB. Respondents A6 and A7 stressed on neutralizing the strong influence of the ruling political party and policy outcome consideration that mostly aims to benefit followers of this party.

Respondent B1 was optimistic that the national tourism policy would have affirmative influences on the Bangladesh tourism industry. Respondent B3 informed that 'we are promoting tourism activities in Bangladesh with policy suggestions'. Respondents B4, B5 and B6 mentioned the thought of forming a 'National Tourism Committee' having the Prime Minister of Bangladesh as the chair. This most authoritative body would coordinate representatives of all governmental ministries and agencies to promote the tourism industry. However, these respondents were not interested to provide any further details regarding this committee. Respondent B2 emphasized on the supportive role of public–private partnerships to encourage more investments towards the tourism industry. Respondent B1 emphasized tourism service development and mentioned that 'the government have plan to create fully functional "One Stop Tourism Service" for the tourists'.

Respondents C1 and C2 had beliefs that carrying out more research studies on tourism and related areas would help. Respondents C3 and C4 added the importance of applying modern research tools to predict the potential tourism growth areas. Respondent C7 supported the opinions of both C3 and C4. Respondent C6 noted that 'the number of domestic tourists is increasing. Still, we cannot see strong efforts to meet their demands'. Respondent C5 offered the considerable point that 'for safety of sea resources and industrial units, if Bangladesh can form the Coast Guard and the Industrial Police Force, respectively, why cannot the government make the 'Tourist Safety Force' stronger as a unit of the Bangladesh Police Force to ensure free movements of tourists across the country?' C6 particularly mentioned the importance of tourists' safety and security, which in recent years has become a major concern in almost all countries. A summary of findings against the theoretical framework is presented in Fig. 6.2. Still, the range of policy suggestions could have included broader, contextual issues such as the climate and global warming, virus attacks, image creation, collaboration, marketing and branding.

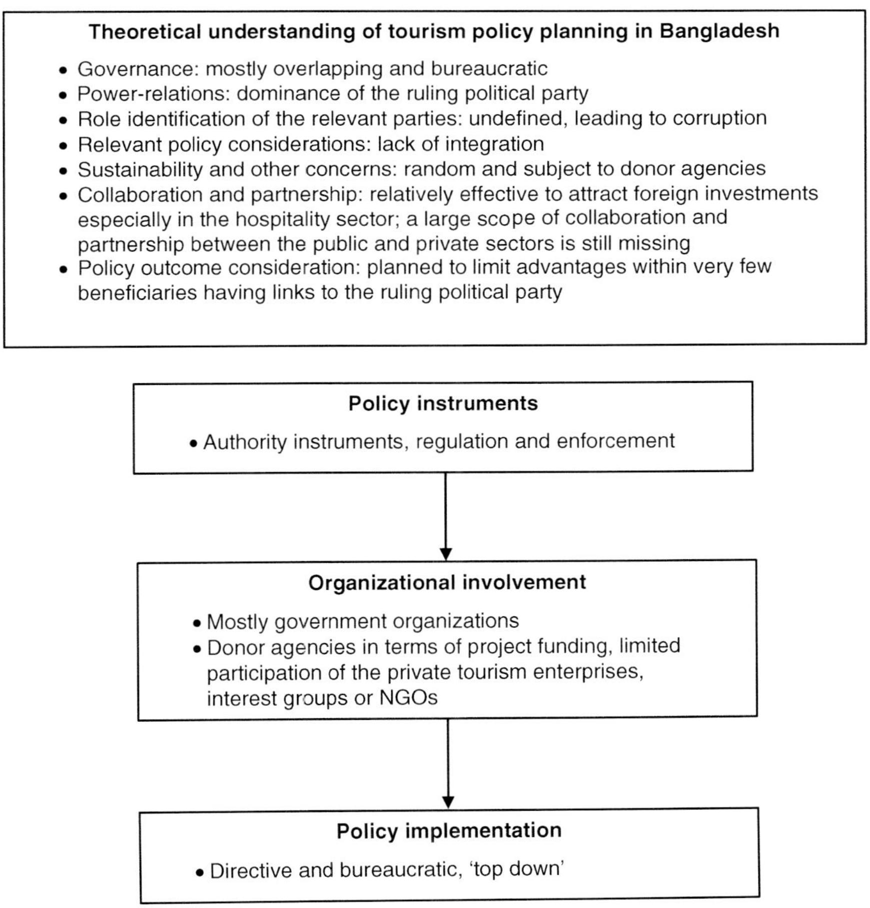

Fig. 6.2. Summary of key findings addressing the theoretical framework (source: author's own elaboration).

Conclusion and Recommendations

The aim of this chapter is to critically discuss tourism policy planning from a comprehensive viewpoint in the Bangladesh context by presenting research having a particular focus on tourism policy planning, instruments, organizational involvement and implementation. Incorporating the respondent views, this research also generates some specific suggestions for tourism policy planning in Bangladesh. Tourism policy planning in a country like Bangladesh needs proper attention from the parties involved for ensuring its effectivity and positive impacts for the beneficiaries. Findings show that tourism policy planning at the national level needs to bring competitive benefits. The tourism industry in Bangladesh still can be defined as at the 'take off' stage that requires sustainable and effective policy planning and implementation (Butler, 2009). Theoretical shifts in the global tourism industry should find place in the national tourism policy of Bangladesh. This research clearly reveals the lack of awareness about the national tourism policy and thus emphasizes the demand for planning and implementing an updated, timely and effective tourism policy. A good number of the general tourist respondents noted their inadequate knowledge about the tourism policy, while the private-sector tourism professionals showed their dissatisfaction. As Telfer and Sharply (2016) argued, that tourism can support in increasing income, foreign exchange, employment, government revenues and boost other subsectors, relevant public-sector tourism policy planners need to have adequate familiarity about the ever-changing global tourism policy dynamics to address ongoing and enduring challenges.

Based on the findings of this research, a number of possible suggestions can be made. First, commitments and coordinated efforts from the government policy planners, private-sector entrepreneurs and general visitors are essential. This should ensure their participation in all stages of the national tourism policy planning. Second, both academic and technical knowledge and the expertise of tourism policy experts and stakeholders need to be incorporated and reflected in the national tourism policy planning. Third, the national tourism policy needs to be planned in line with sustainability concerns as well as contemporary issues to tackle both existing and predicted challenges, where local knowledge needs to be emphasized. Fourth, national tourism policy implementation aspects need to adhere to professionalism and evaluation for ensuring that this can leave positive impacts. Fifth, on the hypothesis of Hall (2009) about 'top down' and 'bottom up' approaches, findings of this research advocate for the 'bottom up' approach where the voices of locals can possibly be heard. Sixth, effective participation of the local people needs to be ensured in the tourism policy and implementation with an aim to apply their 'indigenous knowledge' for the well-being of tourism.

This research could possibly have carried out comprehensive application and testing of indicators of tourism policy planning and evaluation. The lack of literature particularly on the Bangladesh tourism policy was a limitation. Future research studies should cover tourism policy planning aspects from a wider angle based on the lens of findings from this research.

References

Airey, D. and Ruhanen, L. (2014) Tourism policy-planning in Australia: a national and state perspective. *Tourism Planning & Development* 11(2), 149–162.

Alavi, J. and Yasin, M.M. (2000) A systematic approach to tourism policy. *Journal of Business Research* 48(2), 147–156.

Andriotis, K. (2018) *Degrowth in Tourism: Conceptual, Theoretical and Philosophical Issues*. CAB International, Wallingford, UK.

Andriotis, K., Stylidis, D. and Weidenfeld, A. (2018) *Tourism Policy and Planning Implementation: Issues and Challenges*. Routledge, Abingdon, UK.

Asian Development Bank (2016) Sustainable tourism: an opportunity for economic growth in the Pacific. Available at: https://www.adb.org/news/features/sustainable-tourism-opportunity-economic-growth-pacific (accessed 15 July 2018).

BCA (2019) About BCA. Bangladesh Caterers Association. Available at: http://www.bca1960.com/ (accessed 15 July 2019).

Bowen, D., Zubair, S. and Altinay, L. (2016) Politics and tourism destination development: the evolution of power. *Journal of Travel Research* 56(6), 725–743.

BPC (n.d.). National Tourism Policy. Bangladesh Parjatan Corporation. Available at: http://www.parjatan.gov.bd/sites/default/files/files/parjatan.portal.gov.bd/policies/401cea95_b71f_4591_a77d_81eb95e689f6/2020-06-21-12-42-ab744be9913f6906fce79ecee1354d9a.pdf (accessed 14 November 2020).

BTB (2020) Bangladesh Tourism Board. Available at: http://www.tourismboard.gov.bd/ (accessed 13 November 2020).

Butler, R. (2009) Tourism in the future: cycles, waves or wheels? *Futures* 41(6), 346–352.

Chambers, D. and Airey, D. (2001) Tourism policy in Jamaica: a tale of two governments. *Current Issues in Tourism* 4(2–4), 94–104.

Chheang, V. (2009) State and tourism planning: a case study of Cambodia. *TOURISMOS: An International Multidisciplinary Journal of Tourism* 4(1), 63–82.

Collins-Kreiner, N. and Ram, Y. (2020) National tourism strategies during the Covid-19 pandemic. *Annals of Tourism Research.* Available at: https://doi.org/10.1016/j.annals.2020.103076 (accessed 10 March 2020)

Dodd, R. and Butler, R. (2009) Barriers to implementing sustainable tourism policy in mass tourism destinations. *TOURISMOS: An International Multidisciplinary Journal of Tourism* 5(1), 35–53.

Dredge, D. (2006) Policy networks and the local organisation of tourism. *Tourism Management* 27(2), 269–280.

Dredge, D. and Jenkins, J. (2007) *Tourism Planning and Policy*. Wiley, Chichester, UK.

Edgell, D.L. Sr and Swanson, J.R. (2018) *Tourism Policy and Planning: Yesterday, Today, and Tomorrow*. Routledge, Abingdon, UK.

Gunn, C.A. (2002) *Tourism Planning: Basics, Concepts, Cases*. Taylor & Francis, Washington, DC.

Hall, C.M. (2000) *Tourism Planning. Policies, Processes and Relationships*. Prentice Hall, Harlow, UK.

Hall, C.M. (2009) Archetypal approaches to implementation and their implications for tourism policy. *Tourism Recreation Research* 34(3), 235–245.

Hassan, A. and Burns, P. (2014) Tourism policies of Bangladesh – a contextual analysis. *Tourism Planning & Development* 11(4), 463–466.

Hassan, A. and Kokkranikal, J. (2018) Tourism policy planning in Bangladesh: background and some steps forward. *e-Review of Tourism Research (eRTR)* 15(1), 79–87.

Hassan, A., Kennell, J. and Chaperon, S. (2020) Rhetoric and reality in Bangladesh: elite stakeholder perceptions of the implementation of tourism policy. *Tourism Recreation Research* 45(3), 307–322.

Hood, C. (1986) *The Tools of Government*. Chatham House, Chatham, New Jersey.

Inskeep, E. (1991) *Tourism Planning: An Integrated and Sustainable Development Approach*. Van Nostrand Reinhold, New York.

Jamal, T. and Camargo, B.A. (2018) Tourism governance and policy: whither justice? *Tourism Management Perspectives* 25, 205–208.

Kagwanja, P. (2007) Calming the waters: the East African Community and conflict over the Nile resources. *Journal of Eastern African Studies* 1(3), 321–337.

Kennell, J. and Chaperon, S. (2013) Analysis of the UK Government's 2011 tourism policy. *Cultural Trends* 22(3–4), 278–284.

Kilicbeyli, E.H. (2011) Challenges of sustainable tourism development in the developing world: the case of Turkey. *Global Management Journal* 4(1–2), 1–4.

Krutwaysho, O. and Bramwell, B. (2010) Tourism policy implementation and society. *Annals of Tourism Research* 37(3), 670–691.

Lai, K., Li, Y. and Feng, X. (2006) Gap between tourism planning and implementation: a case of China. *Tourism Management* 27(6), 1171–1180.

Lewis, D. (2018) Peopling policy processes? Methodological populism in the Bangladesh health and education sectors. *World Development* 108, 16–27.

Ministry of Civil Aviation and Tourism (n.d.) *Jatiyo Parjatan Nitimala – 2010* [National Tourism Policy – 2010]. Available at: https://mocat.gov.bd/sites/default/files/files/mocat.portal.gov.bd/policies/a7a55206_b33a_4034_b7da_b266c5802553/National%20Tourism%20Circular-2010.pdf (accessed 14 November 2020).

Ministry of Civil Aviation and Tourism (2020) Policies & law. Available at: https://mocat.gov.bd/site/view/policies/Policies (accessed 14 November 2020).

Ritchie, J.R.B. and Crouch, G.I. (2003) *The Competitive Destination: A Sustainable Tourism Perspective*. CAB International, Wallingford, UK.

Selin, S. (1999) Developing a typology of sustainable tourism partnerships. *Journal of Sustainable Tourism* 7(3), 260–273.

Seonyoung, K. and Bramwell, B. (2019) Boundaries and boundary crossing in tourism: a study of policy work for tourism and urban regeneration. *Tourism Management* 75, 78–89.

Stevenson, N., Airey, D. and Miller, G. (2008) Tourism policy planning: the policymakers' perspectives. *Annals of Tourism Research* 35(3), 732–750.

Telfer, D.J. and Hashimoto, A. (2015) Tourism, development and international studies. In: Sharpley, R. and Telfer, D.J. (eds) *Tourism and Development: Concepts and Issues*. Channel View Publications, Bristol, UK, pp. 400–427.

Telfer, D.J. and Sharply, R. (2016) *Tourism and Development in the Developing World*. Routledge, Abingdon, UK.

Tuhin, M.K.W. and Majumder, M.T.H. (2011) An appraisal of tourism industry development in Bangladesh. *European Journal of Business Management* 3(3), 287–299.

UNESCO (2010) State of the World Heritage in Asia and the Pacific 2003 Synthesis Periodic Report for the Asia-Pacific Region. United Nations Educational, Scientific and Cultural Organization. Available at: https://whc.unesco.org/archive/2003/whc03-27com-06ae.pdf (accessed 13 November 2020).

UNWTO (2020) About us. United Nations World Tourism Organization. Available at: https://www.unwto.org/about-us (accessed 13 November 2020).

World Heritage Sites (2020) Bangladesh. Available at: https://www.worldheritagesite.org/country/Bangladesh (accessed 13 November 2020).

WTTC (2020) 2020 Annual research 2019: key highlights. World Travel & Tourism Council. Available at: https://wttc.org/Research/Economic-Impact (accessed 13 November 2020).

Zhang, H.Q., Chong, K. and Ap, J. (1999) An analysis of tourism policy development in modern China. *Tourism Management* 20(4), 471–485.

7 Tourism Development and its Impact on Economic Growth in Pakistan

Sheereen Fauzel, Zameelah Khan Jaffur and Boopen Seetanah*

University of Mauritius, Mauritius

Introduction

In this globalization era, the tourism sector remains one of the largest and most prominent economic sectors in the world. Tourism development has been the primary focus of many recent studies in the literature (Chiu and Yeh, 2017) and the current debate of many economies in the world (Khalil *et al.*, 2007). It is considered as an engine for a country's economic development and growth (Malik *et al.*, 2010; Shahzad *et al.*, 2017). Tourism is believed to spur economic growth in the host country by creating employment opportunities in both the formal and informal sectors, increasing foreign exchange reserves, generating tax revenues, improving the country's balance of payment, developing the country's infrastructure and inducing foreign investment (Khalil *et al.*, 2007; Malik *et al.*, 2010; Ekanayake and Long, 2012; Jalil *et al.*, 2013; Meo *et al.*, 2018).

In the South Asian region, Pakistan is a country which offers a well-diversified list of tourist attractions. It consists some of the most beautiful valleys, lakes, rivers and mountains in the world (Adnan Hye and Ali Khan, 2013). Pakistan is also a country with varied traditions, culture, historical and archaeological sites, and historical civilizations (Khalil *et al.*, 2007; Malik *et al.*, 2010; Arshad *et al.*, 2018). As such, all these unique features make it an ideal adventure, historical and cultural travel destination to visit. Despite this, Pakistan has not been able to attract a large number of tourists to the country until recently since it has been mostly renowned for its tales of terrorism and Taliban rule.[1] Moreover, the tourism industry has been neglected by the few governments in the past and it lacks management that would benefit the tourism industry.

*E-mail: b.seetanah@uom.ac.mu

(eds D. Stylidis and B. Seetanah)
DOI: 10.1079/9781789246698.0007

Since 2010, however, it has been observed that the authorities and various stakeholders (such as tourism establishments and enterprises and tourism employees and professionals) have been more favourable towards the sector (e.g. a branding of the destination,[2] relaxed visa restrictions / new e-visa system and loosening of the No Objection Certificate (NOC) system), supported by the government's willingness to invest in the sector to boost economic growth. Indeed, in recent years the Pakistani government has put tourism at the forefront and centre of its policy agenda. For example, in 2010, the Pakistan Tourism Development Corporation (PTDC) began working with the newly appointed federal Ministry of Tourism to improve the tourism industry. The government also settled the National Tourism Coordination Board (NTCB) to supervise provinces, territories and regions in November 2018. In March 2019, the NTCB redesigned the PTDC to empower provinces and territories to build up their tourism strategies based on their origins. The new visa policy was additionally executed to promote international tourism and the government cancelled the prerequisite of the NOCs to visit a few spots. Another PTDC online interface was likewise propelled to advance the tourism industry and create an information portal for tourists. The Pakistan Tourism Development Endowment finance (PTDEF) of Rs1 billion (Pakistani rupees) was empowered for the advancement and promotion of the cultural heritage of the country and some of the country's elusive destinations. Tourism Promotion and Regional Integration (TPRI) licences were additionally given to promote tourism and local trades. In the same year, the government organized for the first time several national conferences (the Pakistan Tourism Exhibition and Tourism Dialogue, the Pakistan Tourism Summit and the Pakistan Hospitality Exhibition) to support industry relationships between the public and private sectors in order to promote Pakistan as a destination. Rest houses and palatial buildings are being changed over to hotels to give more enchanting facilities to guests (Ghani, 2020).

However, Pakistan's tourism industry still faces certain challenges. For instance, Pakistan's overall performance remains poor as per the Travel and Tourism Competitiveness Index (World Economic Forum (WEF), 2019); it is ranked 121st out of 140 countries (WEF, 2019). And terrorism stigma is still present, although very less, with the country believed to have a decent potential as witnessed by some global followers of the tourism industry. For instance, Pakistan has been declared as the world's third-highest potential adventure destination for 2020 by the British Backpackers Society (BBS) (Ahmed, 2019).[3] This was also supported by an article featured in the *Condé Nast Traveller*, a luxury magazine aiming at providing the best holiday destinations list, in December 2019, whereby Pakistan topped the list for 2020 (Condé Nast Traveller, 2019). In another article, the editors also pointed out that Pakistan is a must-visit country for any adventure traveller. After having listed Pakistan among the ten coolest places to visit in 2019, Forbes recently acknowledged the country as the best travel destination for 2020. In 2010, Pakistan was also termed as being 'tourism's next big thing' in *Lonely Planet: Pakistan* (Bush, 2010)

With around 300,000 tourists in the 1980s, the number of tourist arrivals has overall been on the increase (although not as much as the country would have liked) and reached around 1.9 million in 2018 (Haq, 2018), although encountering a dip in the mid-1980s due to domestic security issues. According to the World Travel & Tourism Council (WTTC), the country's tourism industry directly contributed US$7.91 billion to the country's gross domestic product (GDP) in 2019 (2.87% of GDP) and this is expected to increase to US$16.92 billion by 2029 (2.96% of GDP) (WTTC, 2019). The Pakistani government also supports this and asserts that tourism will be contributing Rs1 trillion (Pakistani rupees) to the country's economy by 2025 (Ali *et al.*, 2019).

Against the above backdrop, the present chapter aims to discuss in the first instance the development of tourism in Pakistan, delving into the government's support of it and its potential as well as the challenges of the industry. The chapter subsequently investigates the impact of tourism development on economic growth in Pakistan over the period 1980 to 2018 using a dynamic time-series analysis, namely a vector autoregressive framework, which allows for the possibility of endogeneity and causality issues. The rest of the chapter is organized as follows. Section 7.1 describes tourism development in Pakistan while Section 7.2 explains the methodology adopted under the present study. Section 7.3 discusses the findings and, finally, a summary and policy implications conclude the chapter.

7.1 Tourism Development in Pakistan

7.1.1 Key tourist attractions in Pakistan

Pakistan is bestowed with an abundance of civilization, tradition, culture, historical sites and beautiful valleys. As an exclusive travel destination, the country has some unique features and offers mainly adventure, cultural and religious experiences, as well as ecotourism. The country offers numerous historical and social sites, landmarks and destinations. Pakistani history, culture and heritage remain key ingredients for the attraction of locals and foreigners to Swat, Kalam, Malam, Jaba, Shangla, Balakot, Ayubia, Murree, Gilgit, Naran, Kaghan and other mountain ranges like the Karakoram, Hindu Kush and Himalayas.[4] Pakistan also has a strong presence in the field of ecotourism and safari tourism. It is worth noting that Pakistan's major urban areas are not left behind, since many foreign tourists also flock to areas such as Lahore, Peshawar, Karachi and Gilgit-Baltistan. These cities have different aspects which portray Pakistan's legacy, culture and traditions. Being the heart of Pakistani culture and the second largest city of Pakistan, Lahore attracts both foreign and local travellers. It also holds a collection of historical artefacts like the Mughal heads and Mughal structures (e.g. the Badshahi Mosque, the tomb of Jahangir, Lahore Fort and Shalimar Gardens) and includes a long rundown (Khalil *et al.*, 2007). The province of Khyber Pakhtunkhwa has a diverse landscape, ranging from mountains, valleys and hills to agricultural farms, and contains several archaeological sites (e.g. Takht Bhai, Pushkalavati, Kanishka Stupa Fort, Bala Hisar Fort among others). It is also well known for its ancestral roots. Taken together, these features make the area a tourist hotspot for adventurers and explorers.

Moreover, the Karakoram Highway, which connects Pakistan to western China, also passes through this area. Karachi, one of the largest cities of Pakistan and that of the country's founder Muhammad Ali Jinnah and his tomb, remains one of the urban areas which is the primary business focus of the country. Besides being home to several excellent and rich hotels (e.g. the Pearl Continental, Marriot, Regent, Sheraton, Carlton, Avari and Ramada Plaza, among others), Karachi is also much admired by tourists for its monuments, buildings and landmarks. Wild rivers and unique landscapes are key features of the city of Gilgit-Baltistan, which attract both domestic and foreign tourists. Indeed, Pakistan is known for its beautiful valleys, plains, mountain peaks (which include three of the world's highest ones), glaciers (among which feature three of the world's seven longest glaciers outside the polar regions – Batura Glacier, Baltoro Glacier and Biafo Glacier), forts, museums and resorts.

7.1.2 Tourism in Pakistan

Since its independence from British rule in 1947, tourism in Pakistan has fluctuated. Tourism became the official interest of the country in the 1960s, when Pakistan was part of the legendary 'Hippie Trail' from Europe to South Asia. It is noteworthy that since 1970 the country has recognized the importance of tourism for the economy, as witnessed by the creation of the PTDC in 1970, which aimed at developing this industry. In the attempt to promote tourism in Pakistan, several institutions and policies have subsequently been developed (Table 7.1).

To further develop the tourism industry in Pakistan, the government established a separate ministry, namely the Ministry of Tourism, in 1996, to cater for tourism in the country. In addition to the coordination and regulation of the tourism activities of both federal and provincial governments and private-sector activities involved in tourism, the Ministry was also responsible for the formulation of policies, development, marketing and promotion of domestic and foreign tourists. The government, in a bid to optimize resources and bring some reforms for enhanced efficiency, transferred the promotion of tourism to the provincial level in 2010. As such, each province started to run its own organization responsible for tourism.[5] These organizations now form part of the PTDC,[6] which is responsible for the development of the tourism sector. Table 7.2 shows the number of tourist arrivals and other key tourism-related indicators for Pakistan over the period 1980–2018.

The tourism industry in Pakistan encountered a downturn throughout the 1980s due to an increase in security issues following the Soviet–Afghan War. During the 1990s, as Pakistan became a hotspot destination for adventure seekers and hiking enthusiasts, the tourism industry of the country started to rise consistently again (PTDC, 2019). Nevertheless, it experienced another decline following the 9/11 attacks in the USA, whereby many Western governments issued severe travel warnings against the country. The decline continued further as the security of the country deteriorated after Pakistan joined the USA in its 'war on terror' to fight against terrorist groups like the Taliban, but shortly

Table 7.1. Institutions and policies to promote tourism in Pakistan (source: authors' own compilation based on data from Pakistan Tourism Forum, 2017; Ministry of Interior, 2019).

Year	Activity
1970	Setting up of the Pakistan Tourism Development Corporation (PTDC)
1972	Creation of the Ministry of Minority, Religious Affairs and Tourism
1976	Establishment of the Department of Tourism Services (DTS); Ministry of Minority, Religious Affairs and Tourism was split and tourism once again became part of the Ministry of Commerce; implementation of Acts of Parliament (The Pakistan Hotels and Restaurants Act 1976; The Pakistan Tourist Guides Act 1976; The Travel Agencies Act 1976) by the DTS
1977	Creation of a separate ministry to account for tourism: Ministry of Culture, Sports, Tourism and Archaeology; approval of several rules with respect to travel agencies and tourism guides
1987	Punjab government established its own Tourism Development Corporation for its province; later, the governments of Sindh and Khyber Pakhtunkhwa also established their own Tourism Development Corporations, for their respective provinces
1990	Presentation of the first National Tourism Policy
1996	Ministry of Culture, Sports, Tourism and Archaeology became the Ministry of Culture, Sports, Tourism and Youth Affairs, subdivided into three parts: Tourism and Sports, Culture and Administration, and Youth Affairs
2000	A Tourism Master Plan was developed under the assistance of the United Nations World Tourism Organization (UNWTO) and United Nations Development Programme (UNDP)
2001	Establishment of the Pakistan-Austrian Institute for Tourism and Hotel Management (PAITHM), Swat, for the promotion of tourism and hospitality education
2004	Tourism was treated in a separate ministry: Ministry of Tourism
2010	Revised draft of the National Tourism Policy was prepared; following the 18th Amendment in the Constitution of Pakistan, the Ministry of Tourism was dissolved, and tourism affairs were transferred to the provincial level
2015	Announcement of the first Provincial Tourism Policy by the government of Khyber Pakhtunkhwa
2019	Adoption of a new visa regime for 175 countries to attract foreigners

after, tourism has been on a constant increase. Under the National Action Plan, two operations, namely the Zarb-e-Azb and Radd-ul-Fasaad, were conducted. Launched in 2014, the operation Zarb-e-Azb was a joint military attack conducted by the Pakistan Armed Forces against various militant groups (such as the Tehrik-i-Taliban Pakistan, the Islamic Movement of Uzbekistan, Al-Qaeda and the Haqqani network). Consequently, an improvement in the overall security situation of Pakistan was noted as terrorist attacks dropped considerably. This operation was followed by another one, the operation Radd-ul-Fasaad, which started in February 2017. The objectives of this operation were threefold: (i) to eliminate the residual threat of terrorism; (ii) to consolidate the gains made by the previous military operation (Zarb-e-Azb); and (iii) to de-weaponize the

Table 7.2. Key tourism indicators in Pakistan, 1980–2018 (source: authors' own compilation based on data extracted from WTTC (2018) and World Bank database).

	1980	1990	2000	2010	2015	2016	2017	2018
Tourist arrivals	292,000	423,800	557,000	907,000	1,247,000	1,756,000	1,750,000	1,800,000
Tourism earnings (US$ million)			551	998	915	791	875	
Direct contribution of travel and tourism to GDP (%)[a]				3.05	2.85	2.88	2.95	2.95
Direct contribution of travel and tourism to employment (thousands)[a]				1,430.0	1,371.2	1,414.6	1,492.8	1,534.2
GDP (US$ billion)		40.01	73.95	177.2	270.6	278.7	304.6	314.6
GDP per capita (US$)	303.50	371.68	519.53	987.41	1356.57	1368.45	1464.99	1482.40
GDP growth (%)	3.8	4.5	4.3	1.6	4.7	5.5	5.6	5.8

[a]Approximate figures.

society. This was of tremendous help to the tourism industry as terrorism and extremism were eliminated from the country. Significant security improvement, coupled with bold government measures with respect to the industry (and the creation of a revised National Tourism Policy in 2010 and its integration into the national development agenda), boosted annual tourist arrivals and after 2010, the industry shot up to the next level. Tourism stood at nearly 1.8 million in 2018. However, there are still some challenges hindering the full potential growth of tourism in the country, mainly related to lack of marketing promotion and branding, infrastructure degradation, air access issues, low information and communications technology readiness, as well as safety and security perceptions among others (WTTC, 2019).

It is noteworthy that a few authors have lately attempted to assess the role of tourism in contributing to the growth of the Pakistan economy. For instance, Khalil *et al.* (2007), Malik *et al.* (2010), Adnan Hye and Ali Khan (2013) and Jalil *et al.* (2013) found the existence of a tourism-led growth hypothesis for Pakistan. Within a simple regression framework, Aleemi and Qureshi (2015) found a significant impact of tourism receipts on economic growth (GDP) over the period 1981 to 2013: on average, tourism receipts contributed about 0.24% in terms of GDP in the case of Pakistan. More recently, Ahad (2016) employed annual data spanning from 1988 to 2014 and examined the tourism-led growth hypothesis for Pakistan: his empirical findings confirmed the existence of a relationship between tourism development and economic growth in both the long and short runs. This chapter uses a dynamic method and a longer time series to reinvestigate the impact of tourism development on economic growth.

7.2 Methodology

In this section, we reinvestigate the impact of tourism development on economic growth for the case of Pakistan using longer and more recent time-series data (1980–2018). Given the dynamic nature of the tourism-led hypothesis, this research employs a dynamic time-series regression analysis, namely a vector error correction model (VECM), which also takes account of both causality and endogenous issues. Grounded on a Solow growth model and previous empirical works from Barro and Lee (1994), Temple (1999), Durbarry (2004) and Seetanah (2011), the following functional model, augmented by a proxy of tourism development, is specified:

$$Y = f(TOU, HC, GDFC, TRADE, FDI) \qquad \text{(Equation 7.1)}$$

7.2.1 Variables measurement and sources

The dependent variable is economic growth (*Y*). Real GDP is used as a proxy to measure economic growth. Various scholars used this proxy, such as Belloumi (2010), Akinboade and Braimoh (2010) and Schubert *et al.* (2011). The variable of interest is tourism development, and this chapter follows the literature on tourism demand and development (Naudé and Saayman, 2005; Durbarry *et al.*, 2009; Seetanah and Khadaroo, 2009; Seetanah and Fauzel, 2019; Fauzel, 2020, among others) by using the number of tourist arrivals (*TOU*) as a measure for tourist development. However, with this proxy, it is not possible to observe every visitor-origin country's economic contribution. Human capital (*HC*) has been overwhelmingly acknowledged to be both a condition for and a consequence of economic growth (Mincer, 1981) as it leads to the production of new knowledge and adapts to innovation and technical changes while increasing productivity. Secondary enrolment ratio is used to measure human capital following Mankiw *et al.* (1992) and Seetanah and Teeroovengadum (2019). The Keynesian school of thought discusses the importance of investment as an instrument to achieve economic growth. Local investment helps to boost output and employment as well as aggregate demand and has a multiplier effect on output (Srinivasan, 2005; Blinder, 2008; Gohou and Soumaré, 2012). Domestic investment (*GDFC*) as measured by gross domestic capital investment as a percentage of GDP is accordingly used as a proxy for local investment. As such the level of trade openness (*TRADE*) is another vector of economic growth, as whenever the economy opens up, there is an increase in exports and foreign currency, with the result of an increase in the welfare of the country (Winters, 2000; Polodoo *et al.*, 2016). The proxy used for trade openness in this study is exports as a percentage of GDP. Finally, foreign direct investment (*FDI*) is also included to capture its impact on economic growth. All data were extracted from the World Bank database (World Bank, 2018).

The econometric specification can be written as follows:

$$\ln GDP_t = \alpha_0 + \beta_1 \ln TOU_t + \beta_2 \ln HC_t + \beta_3 \ln GDFC_t + \beta_4 \ln TRADE_t + \beta_5 \ln FDI_t + \mu_t \quad \text{(Equation 7.2)}$$

where t denotes the time dimension; α_0 represents the intercept; $\beta_i (i = 1–5)$ represents the degree of change in the respective outcome variable for every 1% change in the predictor variable; μ is the error term; and the logarithm is applied to the variables for ease of interpretation and comparison (as the interpretation will be in percentage-change terms).

In addition to the fact that economic growth is a dynamic phenomenon (Seetanah, 2009), the hypothesized relationship may also be endogenous in nature given the possibility of reverse causation, since tourism arrivals and development may also be determined by economic growth. Indeed, such a mutually reinforcing effect is likely to occur, as improvement in economic performance would imply better tourism services and thus yield higher tourism and revenues (Croes, 2012). This chapter adopts a VECM (in the presence of cointegration), which adequately caters for the above while shedding light on both long- and short-term relationships.

7.3 Analysis of Findings

In the first instance, the time-series properties of the data series are tested and results from the augmented Dickey–Fuller (ADF) and the Phillips–Perron tests reveal that all the variables are stationary in their first difference (*I*(1)). Given that the data series are integrated to the same level, the Johansen maximum likelihood approach is used to test for the presence of cointegration. This approach confirms the existence of cointegration, thus concluding the presence of a long-run relationship among the variables. The long-run results after estimating Equation 7.2 using data for the period 1980–2018 are reported in Table 7.3.

The long-run equations provide various interesting results. For instance, referring to the second column of Table 7.3, it can be noted that the coefficient of tourism development is positive and significant and implies that an increase in tourist arrivals in the country contributes to economic growth, thus validating the tourism-led growth hypothesis for the case of Pakistan. Tourism development in this country has had various benefits including employment creation, an inflow of foreign currencies and development of the country's infrastructure as well as improving its development rate. International tourism in Pakistan has achieved record growth. It is noteworthy that the number of tourist arrivals increased beyond 1.7 million in 2017 while tourism revenue reached US$19.4 billion in 2017, representing 6.9% of the country's GDP (WTTC, 2018). As per Wu and Pearce (2014) and Rhaman (2016), the tourism sector in Pakistan is successful and it is expected that it will help in reducing the level of poverty in the country. Moreover, the tourism sector has been successful in boosting the economic growth of Pakistan and similar trends are expected in the future. This result is in line with previous works of Holzner (2011), Seetanah (2011) and Tiwari (2011). Even though Pakistan is a country endowed with a panoply of distinct tourism attractions, based on its unique natural features, biodiversity and rich culture, tradition and heritage, the economic contribution of tourism is relatively still below its potential. This is due to a few challenges related to safety and security, lack of marketing and investment in infrastructure, as well as continued and faster integration of tourism development in Pakistan's development plan.

Referring to the other potential ingredients of growth, it is observed that human capital is crucial for economic growth as per the results reported above (row 3 in Table 7.3). A 1% increase in human capital level increases the

Table 7.3. The long-run estimates of Equation 7.2 for Pakistan, 1980–2018 (source: authors' own computations).

Dependent variable	Ln *GDP*	Ln *TOU*
Ln *GDP*	1.0000	0.7631* (7.5650)
Ln *TOU*	0.3104* (8.2504)	1.0000
Ln *HC*	1.5952* (4.5401)	1.2173** (3.5576)
Ln *TRADE*	0.5913* (2.5109)	−0.4512* (−3.1967)
Ln *GDFC*	1.9711* (2.7381)	1.5041 (0.6952)
Ln *FDI*	0.4351 (0.1008)	0.3320* (4.1583)
Constant	5.0684	3.8679

t-statistics are presented in parentheses.
*Significant at the <10% level.
**Significant at the <5% level.

economic growth of Pakistan by 1.59% as per the present study. This finding confirms the economic proposition on the link between human capital and economic growth. Indeed, a knowledgeable workforce can boost productivity and also improves the quality of work. This result is in line with Sianesi and Van Reenen (2000), Sylwester (2000) and Woolcock and Narayan (2000). It is also observed that international trade is favourable for the country's economy. Improvement in international trade proves to be beneficial for economic growth (row 4 in Table 3) and similar results were obtained by Beck (2002), Katircioglu (2009), Jalil and Feridun (2011) and Shahbaz (2012). Also, this result is in agreement with traditional trade theory and development literature, predicting important growth gains due to trade openness mainly because it results in specialization, investment in innovation, productivity improvement or enhanced resource allocation (Zahonogo, 2016). Another significant observation from the long-run results relates to the local investment coefficient. The results show that an increase in investment results in an increase in national output (row 5 in Table 7.3). This result is in accordance with the endogenous growth model, whereby investment is considered a significant factor that affects economic growth (Epaphra and Massawe, 2016).

Our modelling framework also allows the study of reverse/bidirectional causality. Therefore, we also investigated the impact of economic growth on tourism development and the results are shown in the third column of Table 7.3. It is observed that an increase in economic growth has led to an increase in tourism development, thus confirming a bidirectional causality between these two variables. This result concludes that causality runs through tourism to economic growth and vice versa for the case of Pakistan, and the tourism–growth nexus appears to be a self-reinforced one. The short-run results from the error correction model reveal that the estimated coefficients for tourism and trade are relatively small and even insignificant for the case of investment and education. This implies that growth-enhancing elements take time to have their full effect on economic growth.

Conclusions

This chapter has examined the development of tourism and subsequently empirically assessed the causal effect between tourism development and economic growth for the case of Pakistan over the period 1980 to 2018 using a dynamic time-series analysis. The results proved the existence of the tourism-led growth hypothesis and that tourism development has contributed to economic growth in the long run; perhaps not as much as it could have given the potential of this sector. A bidirectional causality is also reported between tourism development and economic growth, confirming the self-reinforcing nature of the tourism–growth nexus. Moreover, the study also confirmed that human capital, investment and international trade are important ingredients of economic prosperity. The findings also depicted an interrelationship between tourism and two of the variables included in our model (human capital and international trade): an increase in human capital and international trade lead to a rise and a decline in tourism development, respectively. Needless to say, Pakistan is one of the most unexplored countries around the world and remains a hidden natural beauty for many. After securing peace and security in the country, it is high time for the Pakistani government to put the country's tourism industry back on track again. This may necessitate several bold policy measures.

To start with, tourism development should remain one of the very important aspects of the government's vision and should be further integrated into the national development agenda of the country. Hence, it is important for the Pakistani government to prioritize the tourism sector to help increase its performance. The creation of a National Tourism Authority to formulate tourism policies and plans, regulate and establish appropriate coordination with all the key stakeholders of the sector is critical. At the level of provinces, local governments should be involved too, and they will be expected to implement the national tourism plans as well as develop province-specific tourism development plans and policies. Another crucial measure required is the reinforcement of established law and order to maintain social stability in the country. This will improve the image of the country and thus help in increasing tourist arrivals. In terms of infrastructure, policy makers should devote more resources to a better transportation system and high-quality airports in addition to the maintenance of historical and cultural sites.

As highlighted previously, human capital influences both economic growth and tourism development positively. Thus, human resource development in tourism is a fundamental part of the enabling environment for tourism. Indeed, investments in human capital aim at increasing labour productivity and improving the quality of customer service. Research has shown that the quality and competitiveness of the tourist offer directly depend on human capital investments (Folarin *et al.*, 2017). Pakistan should spare no efforts and resources in the formation and upgrading of its human resources related to this sector.

Another important point of this study is the impact of international trade on tourism development. This shows that the country receives fewer revenues from the tourism sector, which is often referred to as invisible exports. As such, to hasten the recent positive trend in tourist arrivals and increase its positive impact on economic growth, Pakistan authorities, together with private-sector operators, should invest heavily (and in a coordinated way) in promoting the destination at both the regional and international levels. The support of travel journalists and travel bloggers, as well as renowned television channels, is crucial in transforming global perceptions of Pakistan. A key role should also be assigned to the embassies abroad in arranging tourism shows to woo the international tourists. Thus, urgent branding or rebranding of Pakistan's identity by relevant policy makers remains key for tourism development and its effect on economic growth.

Although the present study makes several noteworthy contributions to the body of knowledge on the tourism–economic growth nexus in South Asian countries, it is important to recognize one key limitation: data availability. Time-series data on Pakistan are fairly limited. This has restricted the selection of the timespan and variables used in our analysis. Without doubt, large sample size provides more accurate estimates. This study could further be extended by investigating the asymmetric effects of tourism on economic growth through use of non-linear regression models such as the threshold VECM approach or the quantile-on-quantile approach. Use of quantile-on-quantile techniques could help the government to better devise appropriate policies, since it provides an in-depth analysis of the effect of tourism on different quantiles of economic growth. Additionally, each visitor's contribution could be investigated through use of the number of tourist arrivals per country of origin.

Notes

[1] Forming part of the legendary 'Hippie Trail' between Europe and South Asia, Pakistan was welcoming a massive number of foreign tourists from the mid-1950s to late 1970s. After having encountered difficult days in the early 2000s, the country is once again ready to welcome the world's travel community (Ahmed, 2019).

[2] To develop the tourism industry, appropriate measures are required. There are different campaigns like *Amazing Pakistan* and *Visit Pakistan*, which were launched but have not been successful yet in attracting massive tourism.

[3] In 2018, BBS ranked Pakistan as the number one adventure travel destination of the year (Jamal, 2017). The BBS delegation also described the country as 'one of the friendliest countries on earth which offers some of the world's most dramatic mountain scenery' (Associated Press of Pakistan (APP), 2018a). In 2019 during a talk with APP, one of the members of BBS, Samuel Joynson, said that Pakistan is currently 'a travel gem and is one of the world's most exciting travel destinations' (APP, 2018b).

[4] 'Pakistan has more peaks taller than 22,965ft than China and Nepal combined, making it an almost magnetic spot for adventure travellers and intrepid hikers' (Condé Nast Traveller, 2019).

[5] For instance, Punjab Tourism Development Corporation, Sindh Tourism Development Corporation, Khyber Pakhtunkhwa Tourism Development Corporation and Baluchistan Tourism Development Corporation.

[6] The Corporation aims at developing the tourism sector as a national priority in a sustainable and acceptable manner by taking full advantage of regional and international trends and developments so that it can significantly contribute to the improvement of the quality of life in Pakistan and at the same time the promotion of the country's cultural and natural heritage (PTDC, 2019).

References

Adnan Hye, Q.M. and Ali Khan, R.E. (2013) Tourism-led growth hypothesis: a case study of Pakistan. *Asia Pacific Journal of Tourism Research* 18(4), 303–313.

Ahad, M. (2016) Does tourism-led growth hypothesis exist in Pakistan? A fresh look from combine cointegration and causality approach with structural breaks. *International Journal of Economics and Empirical Research* 4(2), 94–111.

Ahmed, S. (2019) Pakistan declared world's third highest potential adventure destination for 2020. *Gulf News*, 29 December 2019. Available at: https://gulfnews.com/world/asia/pakistan/pakistan-declared-worlds-third-highest-potential-adventure-destination-for-2020-1.68714974 (accessed 10 March 2020).

Akinboade, O. and Braimoh, L.A. (2010) International tourism and economic development in South Africa: a Granger causality test. *International Journal of Tourism Research* 12(2), 149–163.

Aleemi, A.R. and Qureshi, M.A. (2015) Tourism receipts and economic growth: empirical evidence from Pakistan. *International Journal of Research* 2(2), 1401–1412.
Ali, R.N., Ehsan, M. and Hassan (2019) Pakistan's tourism – huge untapped potential and way forward. *The Express Tribune*, 28 January 2019. Available at: https://tribune.com.pk/story/1898259/2-pakistans-tourism-huge-untapped-potential-way-forward/ (accessed 5 April 2020).
APP (2018a) British backpackers inspired by Pakistani food, heritage. *PakistanToday*, 17 November 2018. Associated Press of Pakistan. Available at: https://www.pakistantoday.com.pk/2018/11/17/british-backpackers-inspired-by-pakistani-food-heritage/ (accessed 10 March 2020).
APP (2018b) Pakistan to witness a huge surge in international tourism: BBS. *PakistanToday*, 27 March 2018. Associated Press of Pakistan. Available at: https://www.pakistantoday.com.pk/2018/03/27/pakistan-to-witness-a-huge-surge-in-international-tourism-bbs/ (accessed 10 March 2020).
Arshad, M.I., Iqbal, M.A. and Shahbaz, M. (2018) Pakistan tourism industry and challenges: a review. *Asia Pacific Journal of Tourism Research* 23(2), 121–132.
Barro, R.J. and Lee, J.W. (1994) Sources of economic growth. *Carnegie-Rochester Conference Series on Public Policy* 40(1), 1–46.
Beck, T. (2002) Financial development and international trade: is there a link? *Journal of international Economics* 57(1), 107–131.
Belloumi, M. (2010) The relationship between tourism receipts, real effective exchange rate and economic growth in Tunisia. *International Journal of Tourism Research* 12(5), 550–560.
Blinder, A.S. (2008) *Keynesian economics*. The Library of Economics and Liberty. Available at: https://www.econlib.org/library/Enc/KeynesianEconomics.html (accessed 3 March 2020).
Bush, A. (2010) *Pakistan Travel Information and Travel Guide*. Lonely Planet. Available at: https://www.lonelyplanet.com/pakistan (accessed 10 March 2020).
Chiu, Y.B. and Yeh, L.T. (2017) The threshold effects of the tourism-led growth hypothesis: evidence from a cross-sectional model. *Journal of Travel Research* 56(5), 625–637.
Condé Nast Traveller (2019) The best holiday destinations for 2020. Available at: https://www.cntraveller.com/gallery/best-holiday-destinations-2020 (accessed 10 March 2020).
Croes, R. (2012) Assessing tourism development from Sen's capability approach. *Journal of Travel Research* 51(5), 542–554.
Durbarry, R. (2004) Tourism and economic growth: the case of Mauritius. *Tourism Economics* 10(4), 389–401.
Durbarry, R., Nicolas, J.F. and Seetanah, B. (2009) The determinants of tourism demand in South Africa using a dynamic panel data approach. *Tourism Analysis* 14(3), 375–385.
Ekanayake, E.M. and Long, A.E. (2012) Tourism development and economic growth in developing countries. *The International Journal of Business and Finance Research* 6(1), 61–63.
Epaphra, M. and Massawe, J. (2016) Investment and economic growth: an empirical analysis for Tanzania. *Turkish Economic Review* 3(4), 578–609.
Fauzel, S. (2020) The impact of changes in temperature and precipitation on tourists' arrival: an ARDL analysis for the case of a SIDS. *Current Issues in Tourism* 23(10), 2353–2359.
Folarin, O., Oladipupo, E., Ajogbeje, K. and Adeniyi, O. (2017) Does tourism development contribute to human capital development in Africa? *Turizam: međunarodni znanstveno-stručni časopis* [Tourism: an International Interdisciplinary Journal] 65(3), 314–329.
Ghani, S. (2020) 2019 – a successful year for Pakistan's tourism industry. Available at: https://pakistantourntravel.com/2020/01/03/2019-a-successful-year-for-pakistans-tourism-industry/ (accessed 6 April 2020).
Gohou, G. and Soumaré, I. (2012) Does foreign direct investment reduce poverty in Africa and are there regional differences? *World Development* 40(1), 75–95.

Haq, R. (2018) Pakistan's $20 billion tourism industry is booming. Available at: http://www.riazhaq.com/2018/01/pakistans-20-billion-tourism-industry.html (accessed 5 April 2020).
Holzner, M. (2011) Tourism and economic development: the beach disease? *Tourism Management* 32(4), 922–933.
Jalil, A. and Feridun, M. (2011) Impact of financial development on economic growth: empirical evidence from Pakistan. *Journal of the Asia Pacific Economy* 16(1), 71–80.
Jalil, A., Mahmood, T. and Idrees, M. (2013) Tourism–growth nexus in Pakistan: evidence from ARDL bounds tests. *Economic Modelling* 35, 185–191.
Jamal, S. (2017) Pakistan tops list of world's best travel destination for 2018. *Gulf News*, 28 December 2017. Available at: https://gulfnews.com/world/asia/pakistan/pakistan-tops-list-of-worlds-best-travel-destination-for-2018-1.2148655 (accessed 10 March 2020).
Katircioglu, S.T. (2009) Tourism, trade and growth: the case of Cyprus. *Applied Economics* 41(21), 2741–2750.
Khalil, S., Kakar, M.K. and Malik, A. (2007) Role of tourism in economic growth: empirical evidence from Pakistan economy. *The Pakistan Development Review* 46(4), 985–995.
Malik, S., Chaudhry, I.S., Sheikh, M.R. and Farooqi, F.S. (2010) Tourism, economic growth and current account deficit in Pakistan: evidence from co-integration and causal analysis. *European Journal of Economics, Finance and Administrative Sciences* 22, 21–31.
Mankiw, N.G., Romer, D. and Weil, D.N. (1992) A contribution to the empirics of economic growth. *The Quarterly Journal of Economics* 107(2), 407–437.
Meo, M.S., Chowdhury, M.A.F., Shaikh, G.M., Ali, M. and Masood Sheikh, S. (2018) Asymmetric impact of oil prices, exchange rate, and inflation on tourism demand in Pakistan: new evidence from nonlinear ARDL. *Asia Pacific Journal of Tourism Research* 23(4), 408–422.
Mincer, J. (1981) *Human Capital and Economic Growth. NBER Working Paper No. 0803*. National Bureau of Economic Research, Cambridge, Massachusetts.
Ministry of Interior (2019) Policies – Tourist Visa. Pakistan. Available at: https://www.interior.gov.pk/ (accessed 10 March 2020).
Naudé, W.A. and Saayman, A. (2005) Determinants of tourist arrivals in Africa: a panel data regression analysis. *Tourism Economics* 11(3), 365–391.
Pakistan Tourism Forum (2017) *A Position Paper on Challenges Faced by Tourism Industry of Pakistan*. Sustainable Tourism Foundation Pakistan, Islamabad.
Polodoo, V., Seetanah, B. and Sannassee, R.V. (2016) Exchange rate volatility and manufacturing trade: evidence from Africa. *The Journal of Developing Areas* 50(6), 133–148.
PTDC (2019) Pakistan Guide Map. Pakistan Tourism Development Cooperation. Available at: http://www.tourism.gov.pk/publications/pakistan_guide_map.pdf (accessed 1 March 2020).
Rhaman, M.R. (2016) The socio-economic importance of tourism and its impact on the livelihood in South Asia. Case Rangamati, Bangladesh. Bachelor's thesis, Centria University of Applied Sciences, Kokkola, Finland.
Schubert, S.F., Brida, J.G. and Risso, W.A. (2011) The impacts of international tourism demand on economic growth of small economies dependent on tourism. *Tourism Management* 32(2), 377–385.
Seetanah, B. (2009) Is foreign direct investment growth conducive? New evidences from Sub-Saharan African countries, 1980–2005. *Applied Econometrics and International Development* 9(2), 185–201.
Seetanah, B. (2011) Assessing the dynamic economic impact of tourism for island economies. *Annals of Tourism Research* 38(1), 291–308.
Seetanah, B. and Fauzel, S. (2019) Investigating the impact of climate change on the tourism sector: evidence from a sample of island economies. *Tourism Review* 74(2), 194–203.
Seetanah, B. and Khadaroo, J. (2009) An analysis of the relationship between transport capital and tourism development in a dynamic framework. *Tourism Economics* 15(4), 785–802.

Seetanah, B. and Teeroovengadum, V. (2019) Does higher education matter in African economic growth? Evidence from a PVAR approach. *Policy Reviews in Higher Education* 3(2), 125–143.

Shahbaz, M. (2012) Does trade openness affect long run growth? Cointegration, causality and forecast error variance decomposition tests for Pakistan. *Economic Modelling* 29(6), 2325–2339.

Shahzad, S.J.H., Shahbaz, M., Ferrer, R. and Kumar, R.R. (2017) Tourism-led growth hypothesis in the top ten tourist destinations: new evidence using the quantile-on-quantile approach. *Tourism Management* 60, 223–232.

Sianesi, B. and Van Reenen, J. (2000) *The Returns to Education: A Review of the Macro-Economic Literature*. CEE Working Paper. Centre of Economics of Education, London School of Economics and Political Science, London.

Srinivasan, T.N. (2005) Productivity and economic growth in South Asia and China. *The Pakistan Development Review* 44(4), 479–503.

Sylwester, K. (2000) Income inequality, education expenditures, and growth. *Journal of Development Economics* 63(2), 379–398.

Temple, J. (1999) The new growth evidence. *Journal of Economic Literature* 37(1), 112–156.

Tiwari, A.K. (2011) Tourism, exports and FDI as a measure of economic growth: evidence from four Asian countries. *The Romanian Economic Journal* 40, 131–151.

WEF (2019) The Travel & Tourism Competitiveness Report 2019. World Economic Forum. Available at: https://www.weforum.org/reports/the-travel-tourism-competitiveness-report-2019 (accessed 5 April 2020).

Winters, L.A. (2000) *Trade, Trade Policy and Poverty: What are the Links? CEPR Discussion Paper No. 2382*. Centre for Economic Policy Research, London.

Woolcock, M. and Narayan, D. (2000) Social capital: implications for development theory, research, and policy. *The World Bank Research Observer* 15(2), 225–249.

World Bank (2018) World Bank Open Data. Available at: https://data.worldbank.org/ (accessed 12 January 2020).

WTTC (2018) Travel & Tourism Economic Impact 2018: Pakistan. World Travel & Tourism Council. Available at: https://opendata.com.pk/dataset/7a972de2-771f-4f98-bb2b-13717412ade2/resource/eaf03fbc-99f4-4645-8e05-9a3b8c991dc9/download/wttc-pakistan2018.pdf (accessed 17 November 2020).

WTTC (2019) Economic Impact Report data. World Travel & Tourism Council. Available at: https://wttc.org/Research/Economic-Impact (accessed 5 April 2020).

Wu, M.Y. and Pearce, P.L. (2014) Host tourism aspirations as a point of departure for the sustainable livelihoods approach. *Journal of Sustainable Tourism* 22(3), 440–460.

Zahonogo, P. (2016) Trade and economic growth in developing countries: evidence from sub-Saharan Africa. *Journal of African Trade* 3(1–2), 41–56.

8 The Role of International Tourism in the Maldives

Rajasundram Sathiendrakumar[1], Zameelah Khan Jaffur[2] and Boopen Seetanah[2*]

[1]*Sheridan Institute of Higher Education, Australia;* [2]*University of Mauritius, Mauritius*

Introduction

Tourism remains the backbone of the economy of many small island developing states (SIDS) and is often seen as the only means for economic growth in these countries. Indeed, several empirical studies such as Durbarry (2004), Seetanah (2011), Kreishan (2015), Pratt (2015), Cannonier and Burke (2019) and Roudi *et al.* (2019) among others have confirmed that tourism is a key ingredient for economic growth in these countries and that they rely heavily on tourism for the burgeoning of their economies. The developing small island state and South Asian country Maldives, often advertised as a tropical island paradise, is no exception (Henderson, 2008). The country is heavily dependent on tourism, which is the largest economic industry of the islands (Kundur, 2012). Tourism plays a significant role in the economic development of the country by being its largest foreign exchange rate earner and creating employment in its tertiary sector (Sathiendrakumar and Tisdell, 1990; Kundur, 2012; Kapmeier and Gonçalves, 2018). Tourism accounts for about 5–6% of the country's labour force compared with about 35–40% for fisheries in general. Recent statistics also show that tourism's contribution is about 56.6% of the total economy (US$3141.3 million) and 59.6% of the employment (155,600 jobs) in Maldives (World Travel & Tourism Council (WTTC), 2020). Based on pre-COVID-19 estimations, tourism was projected to rise by 5.7% per year and contribute 86.1% of gross domestic product (GDP) in 2028, and tourism employment was expected to rise by 3.1% per year to 47,000 jobs and 19.2% of total employment in 2028 (WTTC, 2018), based on pre-COVID-19 estimations.

The Republic of Maldives is an archipelago consisting of 26 coral atolls stretching 750 km across the Indian Ocean and located on the equator, 700 km

*E-mail: b.seetanah@uom.ac.mu

DOI: 10.1079/9781789246698.0008

south-west of India and Sri Lanka. The atolls contain around 1190 very small islands, of which only 203 are inhabited. Maldives has divided the country into 20 regions formed by merging smaller atolls or by splitting large ones to ease administrative problems. The island of Malé is the capital and represents a separate administrative area. Separate islands are set aside for specific purposes. The Maldives is known for its natural unspoilt environment, such as its white beaches, blue ocean and clean air. Maldivian international tourism entered the world arena in 1972.

From the late 1970s, international tourism has become an important source of foreign exchange for the Maldives. The expansion of international tourism has enabled Maldives to move away from being among the poorest of island developing countries before 2010. In 2012, with a population of more than 300,000, it was a middle-income country with a per capita income of over US$6300. This transformation was due to its ability and the willingness of international agencies to assist Maldives to exploit the economic potential of its pristine marine environment.

In the Maldives, international tourism is synonymous with 'resort islands' (Sathiendrakumar and Tisdell, 1985, 1990). Tourist experience for the Maldives has been marketed based on getting away from 'crowds' and the availability of natural resources unspoilt by human pollution (Sathiendrakumar, 2001). The climate of the Maldives is indeed ideal for visitors to engage in water sports such as swimming, fishing, scuba diving, snorkelling and water-skiing. The natural environment of the Maldives attracts tourists from all over the world every year. Thanks to a series of master plans set by the authorities since 1980 in an attempt to strictly have control on tourism, tourist arrivals in the country expanded considerably from 42,000 in 1980 to over 600,000 in 2004 (Henderson, 2008). Despite that the tourism industry faced some downturns due to the 2004 tsunami which flooded the country, the 2009 global financial crisis (Fitch, 2020) and the state of emergency imposed in 2018 (Maldives Insider, 2019), tourist arrivals have shown an increasing trend in recent years (Fitch, 2020). Maldives attracted a record of 1.7 million of tourists in 2019, achieving a milestone as a tourist destination country and surpassing its target set for the year (1.5 million) and that of 2018 (1.48 million, an increase of around 15%) (Sharma, 2020).

This chapter considers the development of tourism in Maldives and delves into tourism planning and promotion since the 1970s. Despite some studies having been carried out on the tourism sector in Maldives (see e.g. Shakeela *et al.*, 2011; Kundur, 2012; Kapmeier and Gonçalves, 2018; Chia and Muiz, 2020), there is no study to date that has empirically examined the tourism-led growth hypothesis for the case of Maldives. As such, this chapter also empirically investigates the impact of tourism on economic growth in Maldives from 1995 to 2016 in both static and dynamic time-series analysis settings. The rest of the chapter is organized as follows. Section 8.1 provides an in-depth analysis of the evolution of the tourist industry in Maldives and discusses the tourism planning and the various master plans for tourism. Section 8.2 presents the methodology for the present study and empirical analysis of the tourism-led growth hypothesis for the case of Maldives, while the final section consists of some concluding remarks together with some challenges for the Maldivian tourism.

8.1 Evolution of Tourism in Maldives

Table 8.1 shows the evolution of the number of tourist arrivals since 1995 as well as other key tourism performance statistics for the islands. It is noteworthy that the number of arrivals has been consistently on the rise and so has the foreign exchange receipts. The importance of this sector to the economy, which amounts to nearly 60%, is one of the highest around the globe. In terms of employment, recent statistics also show that tourism contributes about 59.6% of the employment (155,600 jobs) in Maldives (WTTC, 2020).

It is interesting to note that tourists bear a tax of 10% on anything which is considered a service facility. Besides, there is a further 12% tourist tax and a US$6 environmental tax, per person per night, on resorts and boats (Mack, n.d.). Money raised through tourism has been invested in healthcare facilities and education centres across the Republic of Maldives, and tourism provides an alternative source of income for some of the young Maldivians who want to move away from the traditional fishing practised by their ancestors and other unsustainable employment in the Maldives (Henderson, 2008).

The regions that supplied a large number of international tourists to the Maldives between 2014 and 2017 are Asia and Pacific, Europe, the Americas, the Middle East and Africa. After a trend of declining European tourists for four years prior to 2017, the year 2017 showed a remarkable recovery, with Europe regaining its position at the top. Prior to 2017, for the previous four years (2013 to 2016), Asia and the Pacific occupied the top spot as the tourist-generating countries to the Maldives. The Americas was the third-largest tourist-generating region to the Republic of Maldives' international tourist numbers. Its market share increased from 2.9% in 2013 to 4.5% in 2017. The Middle East was the fourth-largest tourist-generating region to the Republic of Maldives. The African market was insignificant for the Maldives. It remained almost stable between 2013 and 2017. As at 2019, the greatest share of the arrivals is from Europe, with 833,904 tourists, followed by 705,117 tourists from the Asia-Pacific region (representing a 13.5% increase from the previous year), while the Americas was

Table 8.1. Tourism statistics in Maldives, 1995–2019 (source: authors' own compilation based on data from Trading Economics, 2020).

Year	Number of tourist arrivals	International tourism receipts (US$ millions)	Tourism and travel as % of GDP
1995	315,000	211	44.5
2000	457,000	321	44.8
2005	395,000	826	12.1
2010	792,000	1713	69.8
2015	1,234,000	2691	71.6
2016	1,286,000	2640	65.4
2017	1,390,000	2771	66.1
2018	1,484,000	3054	66.4
2019	1,700,000	3140	57.0

the third major market with 84,793 arrivals, and 60,003 and 18,698 visitors came from the Middle East and African countries, respectively. Most of the tourists who visited the island in terms of a single country came from China (6.7% of total tourists who visited the Maldives in 2019). Although a record number of tourists also visited from Italy, Germany and the UK, Maldives recorded an important increase in tourist arrivals from India in 2019 (166,015 tourists).

8.1.1 Hotels and accommodation

International tourism commenced in 1972 in two resort islands in Kaaf Atoll with a capacity of 280 beds in total. Over time the tourism industry has been expanding in both the north and south of the country. By 2017 (see Table 8.2), the tourism industry had 736 different types of accommodation, comprising four types of establishments: resorts/marinas, hotels, guesthouses and safari vessels (Ministry of Tourism, 2018), licensed and regulated by the parent ministry. Individual resorts/marinas, hotels and guesthouses are located on separate islands and transport is provided by speedboats (Kundur, 2012). On the other hand, safari vessels provide transport directly from the international airport for those staying on them. Packaged agency tour operations are a central element in foreign tourism to the Maldives. Table 8.2 illustrates the numbers of tourist accommodation establishments and their bed capacities from 2013 to 2017. It is noteworthy that hotels usually provide full-board accommodation and are located near the international airport, which itself is located on a separate island.

Resorts/marinas are both an attractive and a luxurious form of accommodation catering for rich people. Each resort is located on one uninhabited island, where locals are not allowed except the people who work in those resorts. Hotels, on the other hand, are located on inhabited islands which are generally closer to the airport while guesthouses were permitted to operate on inhabited islands only in 2009, although secluded resorts have been operating since the early 1970s. They give visitors a cheaper alternative and aim at mid- to low-market segments. These mainly offer bed and breakfast, while some guesthouses are similar to inclusive hotels. One negative impact of having guesthouses on the inhabited islands as opposed to having them on uninhabited islands is the resulting breakdown of the socio-cultural environment for the locals living on those islands. This is because Maldivians have been used to living in small communities undisturbed by foreign tourists and, thus, many of those locals may think that hosting foreigners year-round could destroy their unique way of life. Hence, they may not be likely to welcome the rapid progress of guesthouse businesses on those islands which have been occupied historically by locals. The integration of local communities with guesthouse tourism is thus not an easy endeavour and tourism planners have been working hard on this aspect. Safari vessels are live-aboard floating beds (luxury cruise boats) and mainly attract those tourists who are interested in diving to explore the beautiful corals and/or shipwrecks. They are probably the best way to enjoy this tropical paradise given the ease of mobility

Table 8.2. Type of accommodation and number of beds in Maldives, 2013 to 2017 (source: authors' own compilation based on data from Ministry of Tourism in the Maldives: Document on Tourist Accommodation, 2018).

Type of establishment	2013		2014		2015		2016		2017	
	Number	Beds	Number	Beds	Number	Beds	Number	Beds	Number	Beds
Resorts/marinas	110	23,487	111	24,031	115	24,877	126	26,933	135	28,977
Hotels	19	1,626	19	1,704	17	1,648	16	1,713	10	1,456
Guesthouses	136	1,918	220	3,199	313	4,621	393	6,034	458	7,304
Safari vessels	163	2,716	161	2,739	170	2,839	144	2,694	133	2,518
Total	428	29,747	511	31,673	615	33,985	679	37,374	736	40,255

they provide, and they often possess modern diving equipment and qualified diving instructors.

It is noteworthy that this chain of islands scattered across the equator in the Indian Ocean is grouped into 20 atolls, in order to facilitate the development of tourism in the Maldives. Initially, tourism had been focused closer to the international airport, situated on the island Hulhulé, which is located in Kafu Atoll. As such, 80% of tourist bed capacity was within Kafu Atoll and the four atolls surrounding it. Subsequently, the government policy has been to expand tourism both north and south of the capital Malé, as specified in the various Tourism Master Plans. This resulted in tourism in the Maldives being spread across all parts of the uninhabited islands, amounting to 87 uninhabited islands in the country and accounting for 603 accommodation establishments in the country in 2017 (see Table 8.3 which shows the accommodation facilities by atoll in 2017). This has been accompanied by a few challenges related to co-ordinated regulation as well as environmental pressure.

International tourism in the Maldives is essentially prearranged by tour operators and agents in the tourist-generating countries (Sathiendrakumar and Tisdell, 1989) and major goods and services for maintenance and operation of the resorts are imported from overseas. Thus, the economy of the tourist resorts in the Maldives is separated from the general economy, which relies heavily on the fishing industry. Tourist resorts form enclaves and give rise to a kind of economic and social duality.

8.1.2 Tourism planning and development in Maldives

Tourism in Maldives started back in 1972 and during this time, when the first resort was opened with a capacity of 280 beds, tourism was relatively unplanned and took place according to individual initiatives (Dowling, 2000). Tourism was seen as an opportunity since these early days and 1978 saw the creation of the Department of Tourism and Foreign Investment, which had as one of its mandates to supervise, coordinate and maintain tourism services standards in the sector. As the tourism industry started to grow, the department responsible for tourism development in the country created the Tourism Advisory Board in 1984. The establishment of the Ministry of Tourism in 1988 remains a landmark in tourism development in Maldives and it had the task to administer and develop the industry.

A series of Tourism Master Plans was subsequently drawn up to plan tourism in Maldives and the Ministry of Tourism, Arts and Culture (2013) stated: 'the earliest tourism master plans were about giving guidance to a fledgling industry … blueprints for growth prepared by international experts'. In November 1980, the Department of Tourism and Foreign Investment commissioned Dangroup International under the assistance of the Kuwait Fund for Arab Economic Development to carry out a feasibility study for project preparation in the tourism sector and the outcome was the first tourism plan. The First Tourism Master Plan was presented in 1983 and laid the tactics and strategies for the years 1983 through 1992. It focused on sustainable development of tourism,

Table 8.3. Accommodation facilities in Maldives by atoll, 2017 (source: authors' own compilation based on data from Ministry of Tourism in the Maldives: Document on Tourist Accommodation Facilities by Atolls, 2018).

	Resorts			Hotels			Guesthouses			Total		
Atoll	Number	Beds	% Share of beds	Number	Beds	% Share of beds	Number	Beds	% Share of beds	Number	Beds	% Share of beds
Haa Alifu	3	594	2.0	1	104	7.1	9	87	1.2	13	785	2.0
Haa Dhaalu	0	0	0.0	0	0	0.0	7	62	0.8	7	62	0.2
Shaviyani	2	227	0.8	0	0	0.0	2	18	0.2	4	245	0.6
Noonu	6	1,176	3.9	0	0	0.0	4	40	0.5	10	1,216	3.1
Raa	6	1,616	5.4	0	0	0.0	3	24	0.3	9	1,640	4.2
Baa	13	2,562	8.5	1	148	10.1	31	404	5.5	45	3,112	8.0
Lhaviyani	8	2,250	7.5	0	0	0.0	3	32	0.4	11	2,282	5.9
Kaafu	45	10,094	33.7	6	914	62.8	225	4,523	61.9	276	15,531	40.1
Alifu Alifu	13	2,488	8.3	0	0	0.0	69	712	9.7	82	3,200	8.3
Alifu Dhaalu	17	4,036	13.5	0	0	0.0	44	556	7.6	61	4,582	11.9
Vaavu	2	434	1.4	0	0	0.0	23	260	3.6	25	694	1.8
Meemu	2	400	1.3	0	0	0.0	1	8	0.1	3	408	1.1
Faafu	1	250	0.8	0	0	0.0	6	70	1.0	7	320	0.8
Dhaalu	6	1,512	5.0	0	0	0.0	6	42	0.6	12	1,554	4.0
Thaa	1	152	0.5	0	0	0.0	1	12	0.2	2	164	0.4
Laamu	1	194	0.6	0	0	0.0	5	226	3.1	6	420	1.1
Gaafu Alifu	4	616	2.1	1	136	9.3	6	38	0.5	11	790	2.0
Gaafu Dhaalu	3	550	1.8	0	0	0.0	3	50	0.7	6	600	1.5
Ghaviyani	0	0	0.0	0	0	0.0	2	20	0.3	2	20	0.1
Seenu	2	828	2.8	1	156	10.7	8	120	1.6	11	1,102	2.8
Total	135	29,977	100.0	10	1,458	100.0	458	7,304	100.0	603	38,727	100.0

environmental protection, centralized infrastructure and socio-economic development of the country (Kundur, 2012).

With the assistance of the EU, the Second Tourism Master Plan was presented in 1996 and the central theme of the plan was related to the establishment of policies that were meant to facilitate and foster the private sector. This was a policy shift from the control mindset of the First Plan, with a more important role of the private sector in the Second Plan. The Second Plan focused on six broad strategic directions: (i) balancing regional developments; (ii) easing private-sector investments; (iii) consolidating and strengthening institutions; (iv) developing human resources and cultural aspects; (v) harmonizing tourism and the environment; and (vi) strengthening marketing and promotion.

The Third Tourism Master Plan covered the period 2007–2011 and the policies and strategies incorporated in this tourism plan aligned with policies in the seventh National Development Plan of the country. According to this Master Plan, the vision has been designed to foster and strengthen Maldives tourism as a conductor of economic and social development with environmentally and socially responsible tourism practices. This Master Plan aimed to develop tourism in harmony with nature, facilitating private-sector investment, developing human resources, increasing employment opportunities, and diversifying the market and product. More equitable distribution of tourism benefits across the archipelago was new to the Third Plan. The Fourth Tourism Master Plan of the island state covered the period 2013–2017 and focused on infrastructure enhancement and the creation of special zones for tourism development.

On 30 July 2019, the Fifth Tourism Master Plan was revealed by the tourism minister. The five-year plan laid emphasis on sustainable development of tourism, related to the sustainable use of natural resources and renewable energy as well as environmental conservation, in addition to the conservation of Maldives' heritage and culture, the safety and security of tourists, strengthening the legal framework, as well as providing job opportunities for Maldivians, with special attention to expanding the role of women in the sector.

In addition to the Tourism Master Plans, it is noteworthy that the government has formulated several complementary strategies over time: for instance, a National Biodiversity Strategy; a National Programme for Adaptation to Climate Change (NAPA); a National Energy Policy; three National Environment Action Plans (NEAP) and a Maldives Conservation Programme 2009. Interestingly, several legislations focusing on the sector were enacted since the 1990s with a number of regulations established subsequently, namely the resort regulation, the tourist vessel regulation, the guesthouse regulation, the hotel regulation, the travel agency regulation, the tour guide regulation, the regulation on the protection and conservation of the tourism industry, the lagoon zoning regulation of leased islands for development of tourist resorts, hotels, guesthouses and yacht marinas, and the construction permit regulation. These are updated accordingly as and when required. It is interesting to note that in 2015, the government started to charge US$6 daily as a 'Green Tax' for tourists staying at resorts, hotels and on cruise ships in recognition of the cost of carbon emissions in terms of climate change (Mack, n.d.). Table 8.4 summarizes the key institutional developments and regulations with respect to the tourism sector in Maldives.

Table 8.4. Institutions and regulations for the promotion of tourism in Maldives, 1978 to 2020 (source: authors' own compilation based on data from Kundur and Murthy, 2013; Maldives Inland Revenue Authority, 2015; Ministry of Tourism, 2019; Techera and Cannell-Lunn, 2019).

Year	Activity
1978	Creation of the Department of Tourism and Foreign Investment[a]
1982	The Department was renamed as solely responsible for tourism management[a]
1983	First Tourism Master Plan (1983–1992)[a]
1984	Tourism Advisory Board was set up as a consultative body affiliated to the tourism authority[a]
1988	Establishment of Ministry of Tourism to develop the tourism industry and plan methods of income generation through tourism, to provide guidelines and to administer the industry[a]
1996	Second Tourism Master Plan (1996–2005)[a]
1999	Establishment of Maldives Tourism Act[a]
2006	Establishment of the Maldives Tourism Development Corporation (MTDC), a public entity to help give locals a share in the industry[a]
2007	Third Tourism Master Plan (2007–2013)[a]
2013	Fourth National Tourism Master Plan (2013–2017)[b]
2015	Introduction of Green Tax[c]
2020	Fifth Tourism Master Plan (2020–2025)[d]

[a]From Kundur and Murthy (2013).
[b]From Techera and Cannell-Lunn (2019).
[c]From Maldives Inland Revenue Authority (2015).
[d]From Ministry of Tourism (2019).

Destination marketing and promotion have been identified as an important element in the Maldivian tourism development; the new government recently allocated US$6.7 million for tourism promotion in the state budget for 2019, and US$9.98 million was earmarked for that in 2020. This will not only help in promoting the country as a tourist destination and increase the number of resorts and guesthouses in the islands, to accommodate more tourists, but also will meet the demand of budget travellers who choose guesthouses over luxury resorts (Maldives Insider, 2019).

8.2 Testing the Economic Impact of Tourism for Maldives

8.2.1 Methodology and data

As mentioned previously, tourism is not only the largest foreign exchange earner of the country but also helps to pay for healthcare facilities and education centres through taxes levied and provides an alternative source of income for some youngsters. In an attempt to better understand the impact of the tourism sector on the Maldivian economy, we empirically investigate the tourism–economic growth nexus, often coined the 'tourism-led growth hypothesis', in Maldives (for a global overview of the tourism–economic growth nexus, see Nunkoo *et al.*, 2020). Relying upon the availability of data and building upon existing empirical studies in this area focusing on SIDS and small island economies (e.g. Durbarry, 2004; Singh *et al.*, 2010; Seetanah, 2011; Jackman and Lorde, 2012; Kreishan, 2015; Pratt, 2015; Cannonier and Burke, 2019; Roudi *et al.*, 2019), our econometric model takes the following form after augmenting it with a proxy for labour force participation:

$$Y = f(TOUR, INV, OPEN, H, LAB) \qquad \text{(Equation 8.1)}$$

where *Y*, *TOUR*, *INV*, *OPEN*, *H* and *LAB* stand for economic growth, tourism development, investment, trade openness, human capital and labour force participation, respectively. For ease of interpretation (in terms of percentage change) and comparison, this analysis follows a double-log linear specification. Equation 8.1 is thus written as:

$$\ln Y_t = \beta_0 + \beta_1 \ln TOUR_t + \beta_2 \ln INV_t + \beta_3 \ln OPEN_t + \beta_4 \ln H_t + \beta_5 \ln LAB_t + \varepsilon_t \qquad \text{(Equation 8.2)}$$

where subscript t represents the time index; β_0 and ε_t are the intercept and disturbance terms, respectively; and $\beta_i (i = 1–5)$ are the coefficients of the respective variables. The construction of the variables together with a brief description of each is given in the following paragraph.

The dependent variable, economic growth (*Y*), is proxied by GDP in local currencies in a similar spirit to Kreishan (2015) and Roudi *et al.* (2019). For this chapter, our variable of interest is *TOUR*, which is used to capture the effect of tourism development on the Maldivian economy. For this, we follow existing studies such as Seetanah (2011), Jackman and Lorde (2012), Kreishan (2015) and Pratt (2015) by employing the number of international tourist arrivals. Nevertheless, this proxy does not account for the number of tourist arrivals per country of origin but rather the total number of tourists visiting Maldives. The second and third control variables (*INV* and *OPEN*) are often used for tourism-led growth modelling (Chingarande and Saayman, 2018). Investment (*INV*) is measured by the gross fixed capital formation in local currencies, while trade openness (*OPEN*) by the ratio of trade to GDP (Seetanah, 2011; Cannonier and Burke, 2019). Human capital (*H*) is another key ingredient for economic growth (Mankiw *et al.*, 1992; Ranis *et al.*, 2000; Benos and Zotou, 2014). Due to the unavailability of data on this variable, we employ the literacy rate to capture the effect of human capital on economic growth. Also included is another influential determinant of economic growth: labour force participation (*LAB*). This is proxied by total labour force and it is expected to affect economic growth (Minami, 1964; Bloom *et al.*, 2010; Shahid, 2014). The sample period consists of annual time-series data spanning 22 years (1995–2016). Data were extracted from the World Development Indicators database of the World Bank (World Bank, 2020a).

8.2.2 Empirical results

Before proceeding to estimation of the model (Equation 8.2), the stationary properties of the variables included are checked through the augmented

Dickey–Fuller (ADF) unit root test. The results reveal that all the variables are integrated of order one (*I*(1)), implying that they become stationary after taking first differences. In the first instance, we employ a static regression technique, the ordinary least squares (OLS), to have an immediate picture of the effect of each variable on the Maldivian economy. Clearly, from the empirical results, it can be seen that only three out of the five variables are significant, with each one having a positive effect on the economic growth of the country (see Table 8.5). In particular, a 1% increase in tourism development and human capital cause economic growth to increase by 0.35% and 3.44%, respectively. These results further support those obtained by Durbarry (2004), Seetanah (2011) and Cannonier and Burke (2019) for SIDS. However, in contrast to the study of Shahid (2014), who found that labour force participation does not influence economic growth for the specific case of Pakistan, another South Asian country, our findings depict a positive relationship between labour force and economic growth in Maldives: economic growth increases by 0.29% whenever there is an increase of 1% in labour force. On the other hand, investment and trade openness prove to be insignificant. These are similar to the findings of Cannonier and Burke (2019) within the OLS specification.

The analysis is further extended to investigate the long-run and short-run impacts of tourism development on the Maldivian economy. As such, a dynamic regression specification is considered. Since all of the variables are integrated of order one, the Johansen approach is used to verify whether cointegration exists among the variables included in the model (see Equation 8.2). This approach confirms the presence of cointegration among the variables, thus suggesting that there is a valid long-run relationship among the variables. To this end, in this chapter we opt for the application of a vector error correction model (VECM) for estimation purposes. In addition to taking account of the dynamic behaviour of economic growth (Seetanah and Khadaroo, 2009), this model also caters for causality and endogeneity issues (Marques *et al.*, 2014; Fauzel *et al.*, 2015). After some additional preliminary tests and diagnostics, the labour force (*LAB*) variable is dropped

Table 8.5. OLS estimates of Equation 8.2 for Maldives, 1995–2016 (source: authors' own computations).

Variable	Coefficient
Ln *TOUR*	0.3512*** (4.0486)
Ln *INV*	0.0528 (0.6006)
Ln *OPEN*	0.0004 (0.0458)
Ln *H*	3.4410*** (14.0527)
Ln *LAB*	0.2893* (2.0519)
R^2	0.9890

t-statistics are presented in parentheses.
*Significant at the 10% level.
***Significant at the 1% level.

from Equation 8.2 and the most parsimonious dynamic regression function can be defined as follows:

$$\ln Y_t = \beta_0 + \beta_1 \ln TOUR_t + \beta_2 \ln INV_t + \beta_3 \ln OPEN_t + \beta_4 \ln H_t + \varepsilon_t \text{ (Equation 8.3)}$$

Under the conditions that the variables are integrated of the same order ($I(1)$) and the existence of a long-run relationship among them, Equation 8.3 is estimated within the VECM framework. The results of the long-run dynamics of economic growth are reported in Table 8.6. A glance at the results shows that the direction and significance of the effects, except for the investment variable (*INV*), are similar to those of the OLS estimates. Turning to the estimates, it can be observed that human capital, which is a key ingredient for economic growth, has the largest positive impact on the Maldivian economy: every 1% increase in human capital accounts for about 24% increase in economic growth. This finding corroborates the results of earlier studies focusing on island economies (see Durbarry, 2004; Seetanah, 2011) and the human development–economic growth nexus (see Ranis *et al.*, 2000). Indeed, an increase in human capital is often translated into an improvement in human development (Ranis *et al.*, 2000) and this increases labour productivity (Benos and Zotou, 2014). Referring to the variable of interest (*TOUR*), it can be noted that the coefficient of tourism development is positive and significant. An increase of 1% in tourism development is associated with a 0.92% increase in economic growth. This result is consistent with the positive relationship depicted in the empirical literature and particularly with previous works on SIDS by Seetanah (2011), Kreishan (2015) and Pratt (2015), who employed the same proxy for the measurement of tourism development. Despite being low, this result thus validates the tourism-led growth hypothesis for Maldives, suggesting that an increase in the number of tourist arrivals promotes the economic growth of the country. It is worth noting that tourism development is beneficial to both the country's economy and locals: contributing about 56.6% of the total economy (US$3141.3 million) and 59.6% of the employment (155,600 jobs) as per 2019 annual statistics (WTTC, 2020). Again here, we find that investment and trade openness are insignificant. Both the OLS and VECM estimates are in line with the results obtained by Cannonier

Table 8.6. Long-run estimates of Equation 8.3 for Maldives, 1995–2016 (source: authors' own computations).

Dependent variable: ln *GDP*	
Variable	Coefficient
Ln *TOUR*	0.9178*** (8.5166)
Ln *INV*	−0.1169 (−1.0689)
Ln *OPEN*	0.0101 (1.1361)
Ln *H*	23.6580*** (5.3514)
Constant	98.2993
R^2	0.3039

t-statistics are presented in parentheses.
***Significant at the 1% level.

and Burke (2019), who investigated the tourism–economic growth nexus for a sample of 15 Caribbean countries from 1980 to 2015 under the dynamic system generalized method of moments (GMM) and OLS specifications using similar proxies for investment (gross capital formation) and openness (trade to GDP ratio). They further added that the significance and direction of the effects of investment and trade openness on economic growth are sensitive to the estimation methods used. On the other hand, the variables do not influence the Maldivian economy in the short run.

Conclusions

The contribution of tourism to the economic development of Maldives is undisputable and the government has been sparing no efforts to consolidate and further develop this sector. This has been confirmed through the empirical time-series analysis carried out over the period 1995–2016. Both the static and dynamic regression results depicted a positive and quite sizeable relationship between tourism and the economic growth of Maldives. It is noteworthy that the analysis could not confirm any relationship in the short run, suggesting that tourism development has its full effect on the economy with time. The analysis carried out on the evolution of the tourism industry has also shown that tourism development is beneficial to both the country's economy and locals. Without doubt, tourism planning (though the various Tourism Master Plans) and marketing have played an essential role in Maldives' tourism success.

At the time of writing, the biggest challenge remains the post-COVID-19 recovery, especially given that the most hit sector globally is the travel and tourism sector. Indeed, Maldives has been quite severely hit by the pandemic and, as at end of June 2020, there has been a reported 2337 confirmed cases (4324 per 1 million population) with a death toll of nine persons. In addition to the sanitary hazards, the global lockdown has had an unprecedented effect on the Maldivian economy. Given its overwhelming dependence on tourism, Maldives is among the worst affected countries in Asia, as well as globally, due to the impacts of the COVID-19 pandemic. The World Bank projects the real GDP of the Maldives to contract by 8.5% in 2020 (World Bank, 2020b). The shock to tourism will adversely impact employment and household earnings, given that at least a third of adult males and a quarter of females are engaged in tourism-related jobs. The uncertain global economic outlook also translates into a rather hard time ahead for Maldives.

For the specific case of Maldives, climate and environment changes remain two other key challenges for the development of tourism in the islands, whose economy is tourism-driven. Due to growing tourism, Maldives faces environmental threats (Kapmeier and Gonçalves, 2018), including rising sea levels; over time, this could discourage tourists from visiting the country, bringing about a reduction in tourism revenue. Domroes (2001) has also highlighted that coral reefs have been degraded and flora and fauna threatened due to the high number of tourists visiting Maldives. The sheer amount of solid waste generated by increased tourism is another environmental impact, as disposal of such sewage and waste from the international tourist accommodations into the surrounding sea is likely to damage the corals in the marine environment (Kapmeier and Gonçalves, 2018). Although the above is taken into account within the Tourism Act as from 2006, which stipulates a set of predefined measures for the protection of the environment during construction and the management of sewage and wastewater in an island or place leased for tourism (Techera and Cannell-Lunn, 2019), there is, however, more to be done with respect to environmental protection. There could be a decline in the quality of the marine environment due to this. If this happened, then international tourists who visit the Maldives for scuba diving and also those who come to enjoy the scenic beauty of the Maldives marine environment may not return.

It is noteworthy that during the United Nations Climate Change Conference 2019 (COP25), the Maldivian Minister of Environment highlighted that the country was in a state of climate emergency and requested for immediate international action (see Mohamed, 2019). The minister added that the Maldivian islands are set to face severe beach erosion and flooding with an irreversible negative impact due to global warming. Moreover, 70–90% of Maldives' coral reefs were projected to be destroyed within 10 years if appropriate action is not collectively taken. Despite that significant public-sector investment has been announced for climate adaptation and mitigation, there is also a need for a new Climate Change Act.

Once tourist capacity is fully reached, which may be quite soon, a policy designed to limit further expansion in tourist numbers may be desirable. That is, the Maldives should concentrate its advertising on tourist-generating countries which have high-spending rich tourists, including couples who might be willing to spend their honeymoon and are willing to pay top dollar. Such couples should be attracted through good advertising and marketing efforts by the Ministry of Tourism. Although rich tourists may spend more per day of their stay, the impact of maintaining them will be higher due to higher staff/bed or staff/room ratio compared with other international tourists. But the positive side is that this may lead to an increase in employment potential for locals in Maldives. Furthermore, because of the absence of 'demonstration effect', due to the isolation of locals in tourist resorts except those working at lower levels of the employment chain, the spillover cost from high-spending international tourists may not cause many problems in the Maldives.

To prevent non-sustainable development in the Maldives, the authorities should take adequate responsibility for the assessment of international tourism impacts on environmental and economic factors. Without collecting adequate information, it may not be possible for the government to develop policies that may direct tourism towards sustainability. The strengthening of environmental impact assessments (EIA) to ensure that all the significant impacts associated with new developments are understood remains an important aspect of the above.

Despite that this chapter's analysis remains among the first to empirically investigate the tourism-led growth hypothesis in Maldives, the analysis was limited by the absence of data. Data availability has not only restricted the number of years included in our analysis but also the number of variables considered as potential drivers of economic growth together with tourism development. A further study with more focus on the tourism–economic growth nexus is therefore suggested. To this end, the number of tourist arrivals per country could be employed to examine the contribution of each origin country on the Maldivian economy. Moreover, interaction terms with tourism development could also be used. Since the country is heavily dependent on tourism, these will surely help to devise better policies to mitigate any adverse effect on the economy due to tourism, as is the case due to the COVID-19 pandemic. On the other hand, quantile-on-quantile techniques could also be employed to have an in-depth analysis of the varying effects of tourism across different quantiles of the economic growth of the country.

References

Benos, N. and Zotou, S. (2014) Education and economic growth: a meta-regression analysis. *World Development* 64, 669–689.

Bloom, D.E., Canning, D. and Fink, G. (2010) Implications of population ageing for economic growth. *Oxford Review of Economic Policy* 26(4), 583–612.

Cannonier, C. and Burke, M.G. (2019) The economic growth impact of tourism in small island developing states – evidence from the Caribbean. *Tourism Economics* 25(1), 85–108.

Chia, K.W. and Muiz, A. (2020) The nature and challenges of guest house business in the Maldives: an exploratory study. *Tourism and Hospitality Research.* Available at: https://doi.org/10.1177/1467358420926688 (accessed 26 February 2021)

Chingarande, A. and Saayman, A. (2018) Critical success factors for tourism-led growth. *International Journal of Tourism Research* 20(6), 800–818.

Domroes, M. (2001) Conceptualising state-controlled resort islands for an environment-friendly development of tourism: the Maldivian experience. *Singapore Journal of Tropical Geography* 22(2), 122–137.

Dowling, R. (2000) The Maldives. In: Hall, C.M. and Page, S. (eds) *Tourism in South and Southeast Asia: Issues and Cases*. Butterworth-Heinemann, Oxford, pp. 248–256.

Durbarry, R. (2004) Tourism and economic growth: the case of Mauritius. *Tourism Economics* 10(4), 389–401.

Fauzel, S., Seetanah, B. and Sannasee, R.V. (2015) Productivity spillovers of FDI in the manufacturing sector of Mauritius. Evidence from a dynamic framework. *The Journal of Developing Areas* 49(2), 295–316.

Fitch (2020) Fitch downgrades Maldives to 'B'; outlook revised to negative. *Fitch Ratings*, 30 March 2020. Available at: https://www.fitchratings.com/research/sovereigns/fitch-downgrades-maldives-to-b-outlook-revised-to-negative-30-03-2020 (accessed 3 June 2020).

Henderson, J.C. (2008) The politics of tourism: a perspective from the Maldives. *TOURISMOS: An International Multidisciplinary Journal of Tourism* 3(1), 99–115.

Jackman, M. and Lorde, T. (2012) Examination of economic growth and tourism in Barbados: testing the supply-side hypothesis. *TOURISMOS: An International Multidisciplinary Journal of Tourism* 7(2), 203–215.

Kapmeier, F. and Gonçalves, P. (2018) Wasted paradise? Policies for small island states to manage tourism-driven growth while controlling waste generation: the case of the Maldives. *System Dynamics Review* 34(1–2), 172–221.

Kreishan, F.M. (2015) Empirical study on tourism and economic growth of Bahrain: an ARDL bounds testing approach. *International Journal of Economics and Finance* 7(11), 1–9.

Kundur, S.K. (2012) Development of tourism in Maldives. *International Journal of Scientific and Research Publications* 2(4), 1–5.
Kundur, S.K. and Murthy, K. (2013) Environmental impacts of tourism and management in Maldives. *International Journal of Environmental Sciences* 2(1), 44–50.
Mack, C. (n.d.) Responsible tourism in the Maldives. Available at: https://www.responsiblevacation.com/vacations/maldives/travel-guide/responsible-tourism-in-the-maldives (accessed 5 June 2020).
Maldives Inland Revenue Authority (2015) *Green Tax Guide Version 15.1*. Available at: https://www.mira.gov.mv/forms/Green%20Tax%20Guide%2028-10-2015.pdf (accessed 10 May 2020).
Maldives Insider (2019) Maldives proposes tourism marketing budget hike, allocates $9.98 mln for 2020. *Maldives Insider*, 6 November 2019. Available at: https://maldives.net.mv/33966/ (accessed 3 June 2020).
Mankiw, N.G., Romer, D. and Weil, D.N. (1992) A contribution to the empirics of economic growth. *The Quarterly Journal of Economics* 107(2), 407–437.
Marques, A.C., Fuinhas, J.A. and Menegaki, A.N. (2014) Interactions between electricity generation sources and economic activity in Greece: a VECM approach. *Applied Energy* 132, 34–46.
Minami, R. (1964) Economic growth and labour supply. *Oxford Economic Papers* 16(2), 194–200.
Ministry of Tourism (2018) *Tourism Yearbook 2018*. Ministry of Tourism, Republic of Maldives.
Ministry of Tourism (2019) *Annual Publication 2019 Volume 1*. Available at: https://www.tourism.gov.mv/en/downloads/publications# (accessed 10 May 2020).
Ministry of Tourism, Arts and Culture (2013) *Fourth Tourism Master Plan 2013–2017. Volume 1: Strategic Action Plan*. Available at: https://www.tourism.gov.mv/en/downloads/master_plan (accessed 10 April 2020).
Mohamed, S. (2019) We are living through a climate emergency: Environment Minister. *The Edition*, 11 December 2019. Available at: https://edition.mv/news/13899 (accessed 5 June 2020).
Nunkoo, R., Seetanah, B., Khan Jaffur, Z.R., Moraghen, P.G.W. and Sannassee, R.V. (2020) Tourism and economic growth: a meta-regression analysis. *Journal of Travel Research* 59(3), 404–423.
Pratt, S. (2015) The economic impact of tourism in SIDS. *Annals of Tourism Research* 52, 148–160.
Ranis, G., Stewart, F. and Ramirez, A. (2000) Economic growth and human development. *World Development* 28(2), 197–219.
Roudi, S., Arasli, H. and Akadiri, S.S. (2019) New insights into an old issue – examining the influence of tourism on economic growth: evidence from selected small island developing states. *Current Issues in Tourism* 22(11), 1280–1300.
Sathiendrakumar, R. (2001) The Republic of Maldives. In: Ember, M. and Ember, C.R. (eds) *Countries and Their Cultures, Vol. 3*. Macmillan, New York, pp. 1376–1383.
Sathiendrakumar, R. and Tisdell, C.A. (1985) Tourism and the development of the Maldives. *Massey Journal of Asian and Pacific Business* 1(1), 27–34.
Sathiendrakumar, R. and Tisdell, C.A. (1989) Tourism and the economic development of the Maldives. *Annals of Tourism Research* 16(2), 254–269.
Sathiendrakumar, R. and Tisdell, C.A. (1990) Marine areas as tourist attractions in the Southern Indian Ocean. In: *Proceedings of the 1990 Congress on Coastal and Marine Tourism, Honolulu, Hawaii, USA, 25–31 May 1990, Vol. 1*. National Resources Research & Development Institute, Newport, Oregon, pp. 78–87.
Seetanah, B. (2011) Assessing the dynamic economic impact of tourism for island economies. *Annals of Tourism Research* 38(1), 291–308.

Seetanah, B. and Khadaroo, J. (2009) An analysis of the relationship between transport capital and tourism development in a dynamic framework. *Tourism Economics* 15(4), 785–802.
Shahid, M. (2014) Impact of labour force participation on economic growth in Pakistan. *Journal of Economics and Sustainable Development* 5(11), 89–93.
Shakeela, A., Ruhanen, L. and Breakey, N. (2011) The *Maldives: a sustainable tourism success story? e-Review of Tourism Research* 9(5), 243–263.
Sharma, A. (2020) Coronavirus hits Maldives' lucrative tourism industry. *Deutsche Welle*, 15 April 2020. Available at: https://www.dw.com/en/coronavirus-hits-maldives-lucrative-tourism-industry/a-53131198 (accessed 3 June 2020).
Singh, D.R., Wright, A.S., Hayle, C. and Craigwell, R. (2010) Is the tourism-led growth thesis valid? The case of the Bahamas, Barbados, and Jamaica. *Tourism Analysis* 15(4), 435–445.
Techera, E.J. and Cannell-Lunn, M. (2019) A review of environmental law in Maldives with respect to conservation, biodiversity, fisheries and tourism. *Asia Pacific Journal of Environmental Law* 22(2), 228–256.
Trading Economics (2020) Maldives – International Tourism, Receipts. Available at: https://tradingeconomics.com/maldives/international-tourism-receipts-us-dollar-wb-data.html (accessed 19 May 2020).
World Bank (2020a) World Development Indicators. Available at: http://datatopics.worldbank.org/world-development-indicators/ (accessed 19 May 2020).
World Bank (2020b) The World Bank in Maldives. Overview. Available at: https://www.worldbank.org/en/country/maldives/overview (accessed 10 April 2020).
WTTC (2018) Travel & Tourism Economic Impact Report 2018. World Travel & Tourism Council. Available at https://www.wttc.org/-/media/files/reports/economic-impact-research/regions-2018/world2018.pdf (accessed 10 April 2020).
WTTC (2020) Economic Impact Report data. World Travel & Tourism Council. Available at https://wttc.org/Research/Economic-Impact (accessed 3 June 2020).

Conclusion

Dimitrios Stylidis*

Middlesex University London, UK

This book aimed to address a significant research negligence in tourism planning and development in the South Asian context, by focusing on tourism impacts, development, planning and policy making in this sub-region of the world. The book contains eight case studies of the following Asian countries: India, Sri Lanka, Nepal, Afghanistan, Bhutan, Bangladesh, Pakistan and Maldives. Despite tourism's notable contribution to the economic, socio-cultural and environmental domains, little is known on successful and unsuccessful cases in the South Asian context, along with the issues and challenges tourism planners and developers are facing.

In Chapter 1 of this edited volume, Anu Treesa George, Min Jiang and Terry DeLacy offered a case study on the impacts of community participation and stakeholders' involvement on tourism planning and destination management in Kerala, India. The chapter empirically assessed the contribution community participation in tourism planning makes to the design of responsible tourism initiatives in Kerala via a qualitative methodology incorporating interviews and focus group discussions. The results of this chapter indicate that in the first stages of development, the local community appreciated the tourism industry and its associated benefits including employment opportunities, income generation and increased land value. However, as tourism development started gradually excluding locals from traditional employment and daily activities, this exclusion generated protests and triggered anger against the tourism industry. The findings also suggest that the participatory approach in tourism planning promotes trust between the various parties and supports inclusive development, contributes to additional employment, and increases demand for local products. The case study concludes that the participatory planning approach helps to balance the expectations and needs of the local population, along with those of the tourism industry. Overall, the chapter of

*E-mail: d.stylidis@mdx.ac.uk

 Tourism Planning and Development in South Asia (eds D. Stylidis and B. Seetanah)
DOI: 10.1079/9781789246698.000c

George, Jiang and DeLacy sheds some light on the design and implementation of a holistic plan for sustainable development in the destination.

Another way to handle the negative impacts of tourism, particularly those generated by overdevelopment, is agritourism. As discussed in Chapter 2 of this edited volume, authored by Frank Niranjan and Glen Croy, agritourism has often been adopted in rural areas experiencing challenges to diversify and enhance farm profitability. However, it is also acknowledged that this type of niche tourism requires several resources. For farmers this translates into monetary investment, which is not always feasible. As such, planning and development officials play a crucial role in assessing agritourism's challenges and potential. Niranjan and Croy in their chapter investigated farmers' perceptions of agritourism potential within the South Asian context using a survey distributed to 216 Sri Lankan farmers. Their results indicate farmers' strong interest in agritourism development. Farmers' willingness to engage with agritourism is attributed by the authors to: (i) perceived agritourism's potential contribution to profitability and accessibility to main centres; and (ii) the constraints present in rural settings and the broader absence of tourism products and facilities. Lack of government support was, however, identified by the authors as a major barrier. The chapter contributes to a greater understanding of agritourism planning and development, including farmers' perceptions in emerging economies.

Chapter 3 of this edited volume, by Tek B. Dangi, Birendra KC and Laxman Gautam, reflected on tourism planning and development in Nepal through an analysis of the various country plans, tracing the achievements made and challenges faced in tourism development over the years. The 15 periodic plans produced showcase this South Asian country's commitment to tourism development, which always received high priority from the national government. The chapter shows how plans transformed over the years from initially envisioning an increase in tourist volumes and foreign currency earnings, followed by goals such as employment generation, tourism diversification, market expansion, increasing accessibility, preservation of fragile natural and cultural resources, and increasing tourist length of stay. Dangi, KC and Gautam argue that the presence of a strong institutional structure, achieved via the establishment of the Ministry of Tourism in 1978 as a policy-making body, facilitated the realization of some of these planning goals. They, however, also comment that Nepal still struggles to enjoy its full potential due to a series of obstacles such as ineffective tourism promotion, insufficient accessibility including limited flight connectivity, insufficient and incompetent human resources, fragile institutional framework, and inadequacy of quality products and services. For these reasons, the authors predict that all tourism-related arrivals and revenue targets will remain unmet, including the target of two million tourist arrivals for *Visit Nepal Year 2020*. The authors conclude that a holistic sustainable tourism plan is much needed for Nepal to meet the expected tourism development outcomes.

War and conflicts between different parties have been identified in Chapter 4 of this volume, authored by Abdul Wassay Najimi, as another challenge of tourism development in Afghanistan. This chapter presents a timeline of tourism development in this South Asian country, clearly showcasing how the country and its tourism industry have been affected by external and internal conflicts affecting rural and urban areas. Periods of peace allowed locals to develop their own businesses as local guides or offering accommodation, which were then severely affected by phases of war and violence. Key challenges identified beyond war are the country's insufficient infrastructure, visa issues and high costs for tourists. Tourism can play a critical role in bringing peace and as a key financial resource to develop the post-war Afghanistan. While tourism here is approached as a positive transformative agent, in places like Mexico and its areas it often reinforces violence among criminal groups; especially in sea and sun destinations such as Acapulco and Cancún (Osorio-García and Novo, 2020). The links as such between war, security and tourism should be further studied in different geographical contexts.

In exploring how tourism planning and policy making in Bhutan promote, reinforce or constrain sustainability, Stephen Pratt and Wantanee Suntikul, in Chapter 5 of this edited volume, focused on the areas of food and beverage purchasing decisions, human resources, transportation and souvenir sales in the hospitality industry. The data were collected based on 19 interviews with general managers of hotels in Bhutan. Food and beverages were reported to be the largest source of cost for hotels. Among them, meat is the highest food expense, the vast majority of which is imported. The Bhutanese government's choice not to open an abattoir indicates that meat will be imported in the short run. Obstacles acknowledged in this case study in sourcing locally produced supplies included fluctuation in demand due to seasonality, higher costs as compared with imported goods and a general lack of consistency in quality. Seasonality in human resources was not reported as an issue of concern, as managers responded employing all staff across the year. The vast majority of all hotel employees are locals, although there is a discrepancy in the average monthly salary between local and expatriate employees across all rankings. Although hotels rarely sell souvenirs, those that do sell them prefer the ones locally produced. Overall, the authors conclude that while the controlled nature of tourism in this South Asian country protects locals from the negative impacts of tourism, some of the policies also restrain a wider distribution of tourism benefits. A potential way forward in policy making and development of tourism in Bhutan will involve key changes in the relationships between modernity, tradition, quality of life and sustainability.

Chapter 6 of this edited volume, authored by Azizul Hassan, critically analysed tourism policy planning in Bangladesh. Following a review of the relevant literature, 20 face-to-face interviews were conducted to collect relevant data with: (i) professionals working in private tourism enterprises; (ii) officials in the public sector involved in tourism policy planning; and (iii) domestic tourists. Up to 1992, when the first tourism policy was adopted by the Bangladeshi government, the country had no formal tourism policy but rather a strategic Master Plan for tourism development prepared by the United Nations Development Programme/United Nations World Tourism Organization. The results of Hassan's case study suggest that the policy implementation is characterized as directive and bureaucratic, commonly known as 'top down'. Key challenges identified in the interviews include issues with governance, i.e. mostly overlapping and bureaucratic, while in terms of power relations, there are clear signs of dominance of the governing party. Roles within the policy framework are not always clear and the focus on sustainability is more often than not random rather than the outcome of a plan. There is also room for improvement with regard to collaboration and partnership between the public and private sectors. This chapter contributes to theory and practice by identifying priority areas which benefit effective tourism policy making and assist in accelerating tourism development. Similar to Chapter 1, stakeholders' participation in all stages of the national tourism policy planning has been identified as a key to success. The research concludes that tourism policy making needs to be supportive for the tourism industry and highlights the need to be integrated with the general development policy of a country.

In Chapter 7 of this volume, Sheereen Fauzel, Zameelah Khan Jaffur and Boopen Seetanah identified a series of challenges Pakistan's tourism industry is facing, including the stigma of terrorism. This is well reflected in Pakistan's poor performance ranking as very low in the overall Travel and Tourism Competitiveness Index. The authors followed the history of tourism policy in this country from the initiation of the Pakistan Tourism Development Corporation in 1970 until the adoption of a new visa regime for 175 countries to attract foreigners in 2019. Their chapter focused on the impact of tourism development on economic growth using recent time-series data (1980–2018), by employing a dynamic time-series regression analysis, namely a vector error correction model (VECM). The results of this chapter empirically confirmed the positive impact tourism development has on economic growth in the long run, although not as much as it could have, given the country's and sector's potential. They also indicated an interrelationship between tourism, human capital and international trade. An increase in human capital and international trade lead to a rise and a decline in tourism development, respectively. Human resource development appears critical in facilitating tourism, and investments in human capital can lead in increasing labour productivity and improving the quality of customer service. The authors also conclude that the country should invest in tourism marketing in an attempt to change its image and negative stereotypes related to terrorism and risk.

In Chapter 8, the last chapter of this edited volume, Rajasundram Sathiendrakumar, Zameelah Khan Jaffur and Boopen Seetanah focused on the importance of tourism in small island developing states (SIDS), where the industry is often seen as a key means for economic growth. In their chapter the authors used the case of Maldives, often advertised as a tropical island paradise. Maldives contains around 1190 very small islands, of which only 203 are inhabited. The chapter considered the development of tourism in the Maldives since the 1970s and further investigated the impact of tourism on economic growth from 1995 to 2016, in both static and dynamic time-series analysis settings. Both results illustrated a positive and sizeable relationship between tourism and the economic growth; tourism development is beneficial to both the country's economy and locals. Given its overwhelming dependence on tourism, Maldives is among the worst affected countries in Asia, and globally, by the impacts of the COVID-19 pandemic. Climate and environment changes including rising sea levels remain two other key challenges for the development of tourism. A policy aiming to control tourist numbers is proposed by the authors, shifting the focus to tourist-generating countries with high-spending visitors. The authors concluded that equal attention should be given to policies targeting the protection of the fragile and vulnerable ecosystem in Maldives.

Overall, this edited volume contributes to the documentation and understanding of tourism in the eight South Asian countries by identifying challenges such as terrorism and crime in the case of Afghanistan and Bangladesh, insufficient accessibility and fragile institutional framework, as in the case of Nepal, and overdependence on tourism along with a fragile ecosystem, noted in Maldives. Apart from key challenges, opportunities for tourism development and responses to over-tourism have been documented in this volume, including the case of agritourism in Sri Lanka and the participatory approach of tourism planning observed in Kerala, India. The cases of Bhutan and Pakistan discussed also showcased the importance of tourism, calling though for additional planning and more effective governance to sustain its progress. All cases have established the role of tourism as an agent of change, with its success, however, depending to a large extent on public planning and policy making at the national, regional and local levels.

While this book is among the very few attempts to document tourism development and planning in South Asian countries, additional research is deemed necessary. Although all eight countries of this sub-region have been included in this book, the size and complex context of some countries, like India and Bangladesh, call for additional contributions from these areas. Second, this book responds to previous research deficit by focusing on tourism development and planning in South Asian countries; however, the fact that each chapter deals with a different area does not allow for comparisons to be made. Particular tourism development issues in each country might have been neglected, and it was not possible to identify generalized patterns of development in the sub-region. Key lessons, however, can be drawn from the presence of tourism in this sub-region, useful to planners, developers and policy makers in identifying ways to reduce some of the issues and increase the benefits.

Reference

Osorio-García, M. and Novo, G. (2020) Tourism development in Mexico. In: Monterrubio, C., Andriotis, K. and Stylidis, D. (eds) *Tourism Planning and Development in Latin America.* CAB International, Wallingford, UK, pp. 104–122.

Index

Note: The locators in bold and italics represents tables and figures respectively.

This book is published by **CABI**, an international not-for-profit organisation that improves people's lives worldwide by providing information and applying scientific expertise to solve problems in agriculture and the environment.

CABI is also a global publisher producing key scientific publications, including world renowned databases, as well as compendia, books, ebooks and full text electronic resources. We publish content in a wide range of subject areas including: agriculture and crop science / animal and veterinary sciences / ecology and conservation / environmental science / horticulture and plant sciences / human health, food science and nutrition / international development / leisure and tourism.

The profits from CABI's publishing activities enable us to work with farming communities around the world, supporting them as they battle with poor soil, invasive species and pests and diseases, to improve their livelihoods and help provide food for an ever growing population.

CABI is an international intergovernmental organisation, and we gratefully acknowledge the core financial support from our member countries (and lead agencies) including:

UKaid from the British people

Ministry of Agriculture People's Republic of China

Australian Government Australian Centre for International Agricultural Research

Agriculture and Agri-Food Canada

Ministry of Foreign Affairs of the Netherlands

Schweizerische Eidgenossenschaft Confédération suisse Confederazione Svizzera Confederaziun svizra Swiss Agency for Development and Cooperation SDC

Discover more

To read more about CABI's work, please visit: **www.cabi.org**

Browse our books at: **www.cabi.org/bookshop**,
or explore our online products at: **www.cabi.org/publishing-products**

Interested in writing for CABI? Find our author guidelines here:
www.cabi.org/publishing-products/information-for-authors/